"Gilles Sabourin succeeds masterfully in bringing to life this exciting period of nuclear physics research, with an original focus on the important and often overlooked contributions of the Canadian-Franco-British group at Montreal. With many previously unknown personal anecdotes and scientific details this book reads as a thriller for the lay reader while offering novel insights to historians. The special circumstances of an international group of world-renowned experts (including my father) striving to unravel the secrets of the atom during the war has unexpected but fascinating parallels with today's efforts to understand novel viruses and develop vaccines in record time. The mutual suspicion and lack of confidence between scientists, politicians, and the general public combined with the rhetoric of those striving for national dominance sounds all too familiar today. We have a lot to learn from this book to avoid repeating the mistakes of the past and to ensure that scientists are offered the moral and financial support that will allow them to work securely and effectively for the good of mankind."

Philippe Halban, Emeritus Professor of Medicine, University of Geneva, son of Hans Halban who headed the Montreal Lab.

"I was pleased to read Gilles' book on the Manhattan Project during WW2 in Montreal ... this less spectacular branch of the nuclear programme has received very little recognition and is largely forgotten. In fact much of the work done in Montreal forms the basis of many medical and industrial applications and I feel privileged to have been part of it."

Alma Chackett, Chemist employed by the Tube Alloy project and wife of Dr. Ken Chackett.

MONTREAL **AND THE BOMB**

Gilles Sabourin

MONTREAL
AND THE BOMB

Translated from the French by Katherine Hastings

Baraka
Books

Montréal

© Baraka Books

ISBN 978-1-77186-265-3 pbk; 978-1-77186-266-0 epub; 978-1-77186-267-7 pdf

Cover by Maison 1608
Book Design by Folio infographie
Translated from the French by Katherine Hastings
Editing and proofreading by Robin Philpot, Rachel Hewitt, Blossom Thom

Legal Deposit, 4th quarter 2021
Bibliothèque et Archives nationales du Québec
Library and Archives Canada

Originally published under the title *Montréal et la Bombe* © Septentrion;
publié avec l'autorisation de Les Éditions du Septentrion

Published by Baraka Books of Montreal

Printed and bound in Quebec

TRADE DISTRIBUTION & RETURNS
Canada – UTP Distribution: UTPdistribution.com

UNITED STATES
Independent Publishers Group: IPGbook.com

We acknowledge the support from the Société de développement des entreprises culturelles (SODEC) and the Government of Quebec tax credit for book publishing administered by SODEC.

Société
de développement
des entreprises
culturelles
Québec

Funded by the Government of Canada
Financé par le gouvernement du Canada | Canada

CONTENTS

FOREWORD

This book is the end result of a long and fascinating research journey. For years I have been gathering everything I can get my hands on about the Montreal Laboratory and the people who worked there, unearthing information on over 400 of the Laboratory's employees in the process. I managed to track down a handful of people still alive today who worked at the Montreal Laboratory during World War II. I also had the opportunity to meet and interview two of them—chemist Alma Chackett and "computer" Joan Wilkie-Heal—and reached out to some thirty individuals whose parents once worked at the Montreal Laboratory. All of these people were extremely generous in sharing with me a huge quantity of documents and photos, as well as their memories about the day-to-day work at the Laboratory. The diary belonging to Hans Halban, the Laboratory's first director, which was graciously provided by his son Philippe Halban, was especially helpful.

I believe the fact I work in the nuclear sector as an engineer specializing in power plant safety has provided me an insider's perspective and the relevant knowledge to fully appreciate the work carried out in Montreal during the war.

I leaned heavily on both the British and Canadian national archives, both of which contain a wealth of information about the Montreal Laboratory. I read and drew from many of the 600 reports produced by the project during the war that are archived in London's Kew Gardens and available in electronic format. The books listed in the bibliography were also of great assistance.

It was these sources, together, that enabled me to reconstruct the story that follows.

INTRODUCTION

Just like every other evening, William Lyon Mackenzie King, Canada's prime minister, headed to his office to dictate his diary entry to his secretary. It was a habit he had maintained for decades, even in the thick of World War II. The entry from that day—August 6, 1945—would occupy several typed pages. First, he would relate the events of the day, which began typically enough with matters of domestic policy but, more importantly, he would recount a crucial moment that was to mark a turning point in history.

Mackenzie King had learned via ministerial memo that the first atomic bomb had been dropped on a Japanese city. Once the shock of the news wore off, the prime minister of Canada, a country aligned with the Allies, evoked the strange sense of hope generated by the explosion: "Naturally, this word created mixed feelings in my mind and heart. We were now within sight of the end of the war with Japan." Mackenzie King ardently hoped that this massive destruction of human life would be the last, and that it would bring an end, once and for all, to the painful conflict between Japan and Canada that began in 1941. He pondered with horror what might have been the outcome had it been the German scientists who won the race to develop the bomb, suggesting the "British race" would have been entirely wiped out.

The prime minister also acknowledged in his diary that day the scientific accomplishment the Americans had just achieved. After all, he knew only too well how difficult it was to tame the atom. For the past two years, his country had been home to a laboratory hidden away in the city of Montreal that had been devoted to that very challenge. While Mackenzie King noted with satisfaction in his diary how extraordinary it was that the nuclear project had been kept secret during that time, he nonetheless expressed his apprehension about certain Allies: "I am a little concerned about how Russia may feel, not having been told anything of this invention or of what the British and the U.S. were doing in the way of exploring and developing the process."

Although Mackenzie King was certainly aware of the stakes surrounding the atomic bomb, thanks to its ramifications in Montreal, he had no knowledge of certain episodes that had played out in his own country. And Montreal's atomic adventure had more than its share of drama...

ORIGINS OF THE LABORATORY

Arrival of Hans Halban in Montreal

For a man arriving in Montreal in 1942 for the purpose of working on a British-led project to develop an atomic bomb, it would be in his interest not to be mistaken for a Nazi! That was surely something front of mind for Hans von Halban, the first man to direct this monumental project. With his German-sounding name and slight German accent, the physicist must have bitterly regretted the decision made by his grandfather, a senior bureaucrat in the Austro-Hungarian Empire who, at the turn of the century, had altered the family's surname. The "von" bestowed on him by the Emperor himself was reserved for nobility, and reinforced the name's "Germanness." In an irony of sorts, several decades later, his grandson would drop the nobiliary particle in an attempt to obscure his roots as an Austrian noble. When Hans Halban arrived in Canada's largest city in early November of 1942, he was keenly aware of the challenges that lay ahead.

Halban was a man of average height, with a high forehead and piercing gaze. He was charming, self-assured, and perfectly at ease in social situations. Born in Germany in 1908, he spoke English and French fluently. These were all traits he would need to draw on to carry out his perilous mission: Hans Halban's role was to build a nuclear physics laboratory—from the ground up—in Montreal. He had time to ponder the project during his trans-Atlantic crossing on a low-altitude cruiser[1] (Halban had a heart malformation that prevented him from flying in a conventional aircraft at high altitudes). This privileged mode of transportation suggested Halban's importance in the eyes of the British, who had pinned their considerable hopes on the renowned physicist. With their island under threat from the Nazis, the British had just tasked Halban with relocating their Cambridge-based nuclear laboratory to Canada. The research being carried out there was highly advanced and was key to the country's military strategy. Whether as a source of energy or a force for destruction, all signs pointed to the atom playing a crucial role in the outcome of the war.

Upon his arrival, the scientist settled in at the Windsor Hotel, one of Montreal's most prestigious establishments.[2] The Windsor was where King George VI and his wife Elizabeth stayed during their visit in 1939.[3] It was also there that General de Gaulle would give a speech before a huge crowd gathered in Dorchester Square in July 1944.[4] At the time, the hotel was located in a rapidly changing neighbourhood that was to become the heart of the city's downtown. Just across the square stood the Sun Life Insurance Company's skyscraper, the tallest building in the British Empire at the time. Hans Halban was soon joined by his wife and his three-year-old daughter Catherine Mauld. While at the hotel, he began meeting with potential candidates to build his team in Montreal. Now all they needed was to find somewhere to conduct their extensive scientific experiments.

Hans Halban, the first director of the Montreal Laboratory, was a scientist born into Austrian nobility who later became a French citizen. He is pictured here in London after the war. It was Halban, together with his colleague Kowarski, who fled in 1940 ahead of the Nazi invasion along with the supply of heavy water that eventually ended up in Montreal. (Lotte Meitner-Graf, personal archives of Philippe Halban.)

It was McGill University principal Frank Cyril James who stepped forward to help find the laboratory a home.[5] McGill was already actively involved in a number of military projects, including one to synthesize RDX, an explosive more powerful than TNT. Frank James was keenly aware of the stakes of the war, and offered Halban the use of an imposing, two-storey home in the Richardsonian Romanesque style, so named after celebrated American architect Henry Hudson Richardson,[6] and inspired by Europe's churches. The stately home boasted a tower and several dormer windows[7] set into grand vaulted arches. The ground floor featured several large rooms with fireplaces, while the upper floors housed a series of bedrooms. Halban took the master bedroom for himself, relegating his secretary to the bathroom. The house stood at 3470 Simpson Street,[8] on the flanks of Mount Royal, the hill dominating the city's skyline. It was extremely convenient for Halban, as it was close to his hotel.

The choice of the McGill building was quickly approved as a temporary solution until a larger and more appropriate location could be found. In 1942, the English-language university had only one other rival in the city—Université de Montréal, a former satellite of Quebec City-based Université Laval. However, McGill had an irrefutable asset: an excellent international reputation for physics. It owed this considerable advantage to New Zealander Ernest Rutherford, a professor at the university from 1898 to 1907. Upon arriving in the city, Rutherford devoted his studies to radioactivity, a discovery made two years earlier by Henri Becquerel and confirmed by Marie Curie. Radioactivity is the spontaneous emission of radiation from matter. Researchers soon realized that several types of radiation were emitted. At McGill, Ernest Rutherford sought to characterize them, working in collaboration—and in competition—with European research teams. His work earned him the 1908 Nobel Prize in Chemistry, and with that distinction, Montreal's nuclear reputation was born!

The British team's atomic laboratory was hardly out of keeping with the local industrial landscape at the time. Montreal was at the heart of the British Empire's military production, and a considerable amount of military equipment was being manufactured in the city. When Canada entered the war alongside England in 1939, it gave Montreal's employment rate a massive boost. Several munitions and armaments factories began operating at full capacity. Defence Industries Limited (DIL) refurbished and restarted a World War I munitions factory in Verdun,[9] a neighbourhood of Montreal, making it the largest manufacturing facility of its kind in Canada. At the height of the war, it employed 6,000 people, most of them women, working in 11-hour shifts. The DIL in Verdun manufactured some billion and a half small-calibre shells that it sold to the Canadian Army and its allies.[10] Tanks rolled off the assembly lines at the Angus Shops in the east end of the city, and destroyers were built by Canadian Vickers in Viauville, another nearby neighbourhood.

Despite all the activity, the researchers recruited by Halban discovered a much different atmosphere in Quebec from the one in Europe. Montreal was spared the ravages of war. There were no German bombers buzzing the skies over Canada. Like other big North American cities, Montreal prepared nonetheless for such an eventuality, albeit somewhat casually. Blackout drills were not always taken seriously, as illustrated by the blackout simulation one night in October 1943, when all the lights at the Central Train Station and McGill University were left on.[11] Rationing was much less

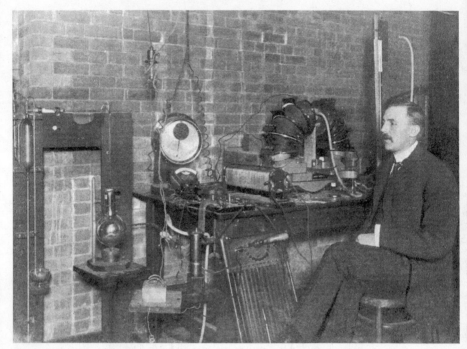

Physicist Ernest Rutherford (pictured here in his laboratory in 1905) won the Nobel Prize in Chemistry for the work he conducted at McGill University. (Photographer unknown, published in A.S. Eve, *Rutherford: Being the Life and Letters of the Rt. Hon. Lord Rutherford, O.M.*, Cambridge University Press, 2013.)

severe than in England, and the British expats who followed Halban to Canada immediately begin sending their families back home packages containing all manner of things, including butter, which had become a rare commodity on the other side of the Atlantic. So much so that Canada Post began offering leakproof boxes to prevent melting butter from oozing while in transit.[12]

Hans Halban was happy to find himself in the relative serenity of the streets of Montreal, after a series of hurried departures: The physicist had been forced to leave France, then England, not only to flee the Nazis, but also to protect a very special substance that would play a vital role in the work of the Montreal Laboratory.

The Battle for Heavy Water

Water–or H_2O–is without a doubt the most widely recognized chemical formula. But what is less common knowledge is that it can also exist in another form, known as heavy water. In this configuration, the hydrogen symbolized by "H" contains both a neutron and a proton in its nucleus. This makes it heavier and, more importantly, makes it useful for nuclear reactions. When a uranium nucleus splits into two, it releases energy and neutrons. If the neutrons are channelled during this process known as "fission," they can trigger other reactions of the same kind, creating the renowned chain reaction used to power nuclear energy plants or explode bombs. One of the men behind this fundamental discovery was none other than the Laboratory's head physicist, Hans Halban. Little did he know, as he was conducting these experiments in the early months of 1940, that he would soon be setting out for Montreal. His career in France was well established: After working with the brilliant Niels Bohr in Copenhagen, he found himself in the holy of holies—the laboratory founded by Marie Curie in Paris. Together with Frédéric Joliot, who was married to Curie's daughter Irène Joliot-Curie, they identified heavy water as an ideal substance to use in nuclear reactions and harness the precious energy they produced.

The only problem was that heavy water is highly rare in nature, and only one of its particles is contained in every 3,200 molecules of ordinary light water! In 1939, heavy water

The Vemork power plant in Norway played an important role in the early days of World War II. This was where the French Secret Service went to collect heavy water in 1940, thereby depriving the Germans of a precious resource. (Anders Beer Wilse, National Library of Norway.)

was still a rarity. Only one plant in the entire world made it in any appreciable quantity—the Norwegian company Norsk Hydro, at its Vemork hydroelectric plant west of Oslo. Heavy water was actually merely a sideline for the plant, which was devoted primarily to making ammonia, for use in fertilizer production. The heavy water produced as a by-product was sold to physicists and chemists for pure research purposes. In early 1940, Frédéric Joliot was well aware of the strategic value of the liquid, and he alerted the French authorities, urging them to get their hands on the entire stock of heavy water at Vemork, amounting to about 185 kilograms. Two French secret agents subsequently made their way to Norway, which at that time was in neutral territory, and convinced the general manager of Norsk Hydro, Axel Aubert, to lend all the heavy water in his possession to France for the duration of the war. The French then secretly transported the barrels of heavy water by plane from Oslo to Perth, Scotland, then on to Paris in early March 1940. It was only the beginning of a long voyage for those seemingly innocuous barrels. But it was the eleventh hour for the French as, on April 9, German troops invaded Denmark and Norway.[13]

Hans Halban and his colleague Lew Kowarski got to work as soon as the heavy water arrived at the Curie Laboratory. The collaboration between the two men, which would continue on to Montreal, was strained. However, they managed to set aside their differences and focused on their experiments to measure the average number of neutrons produced through fission when uranium rods were immersed in heavy water. That number was crucial in determining the masses of uranium and heavy water required to obtain a self-sustained reaction. But their work was soon interrupted by the advance of the Wehrmacht, the army of the Third Reich.

After crossing Belgium, German troops invaded France on May 13, 1940, descending on Paris a month later. In late May, the French Armaments Minister, Raoul Dautry, telephoned Joliot, urging him to relocate his project outside Paris. It was thought at that time that the German forces could be contained north of the Loire River. So, Joliot rented a villa in Clermont-Ferrand, 400 kilometres south of Paris, where he planned to set up a makeshift lab. He tasked his collaborators, including Halban and Kowarski, with transporting the heavy water and all their laboratory notes there. Halban was the first to arrive in Clermont-Ferrand. Kowarski followed in early June at the head of a convoy transporting several tonnes of uranium oxide.

The team made up of physicists Lew Kowarski, Frédéric Joliot, and Hans Halban was at the forefront of atomic research in 1940. (Jules Guéron, courtesy of Emilio Segrè Visual Archives, American Institute of Physics.)

Kowarski joined Halban, while the convoy of uranium headed for the port city of Bordeaux. The French didn't want the Germans to get their hands on the material in their possession. So, when the barrels of heavy water arrived in Clermont-Ferrand, Joliot had them secured in a prison in Riom, several kilometres away. As for the uranium, it was hidden away for the duration of the war and, alas, the researchers were unable to access it. As we will see, the mineral would be in desperately short supply in the months to come.

The Joliot-Curie couple and their two children arrived in mid-June. A temporary laboratory had already been set up in Clermont-Ferrand, and the physicists planned to continue their research there. On June 16, two days after the fall of Paris, the Joliot-Curies were enjoying their breakfast at a café, when a Simca motorcar roared into the square and out stepped Lieutenant Jacques Allier.[14] He asked to speak to Frédéric and took him aside. The French and British armies were in disarray, and the government had just ordered that the heavy water be transferred to Bordeaux, where it was to be loaded onto a ship bound for England. The lieutenant handed Joliot the order signed by Minister Dautry. That same day, Prime Minister Paul Reynaud submitted his resignation to the president of the Republic. He was replaced by Marshal Pétain. The evacuation of the heavy water to England was one of the final decisions of the Reynaud Government. Joliot was convinced that Halban, Kowarski, and their families should leave immediately for Bordeaux, some 500 kilometres away, but he was undecided about his own plans. His wife was unwell. Afflicted with tuberculosis since the early 1930s, Irène Joliot-Curie had already been admitted to a sanatorium on several occasions. Frédéric was torn between continuing his research overseas and remaining in France.

The events that followed were like something straight out of an adventure movie. The next day, June 17, Lieutenant Allier and Halban went to collect the barrels of heavy water from the prison in Riom, not without difficulty, as the prison warden was reluctant to hand them over now that the government had resigned. Allier had to threaten the man with a drawn revolver before he would give them up.[15] Once back in Clermont-Ferrand, Halban and Kowarski packed up their families, their luggage, and their precious heavy water.

It was a long journey, as they had to cross numerous roads running north/south that were clogged with thousands of people fleeing the German-occupied zones. Their convoy arrived in the middle of the night in Bordeaux, where

The dedicated work of Irène and Frédéric Joliot-Curie on radioactivity earned them the Nobel Prize in Chemistry in 1935. (Meurisse Press Agency, Bibliothèque nationale de France.)

one of Allier's aides was awaiting them with hastily scribbled boarding papers for the *SS Broompark* that was docked in the port of Bordeaux.

They loaded all their belongings on board the steamship. Since the few available cabins were set aside for the women and children, Halban and Kowarski had to content themselves with sleeping on a heap of coal. Meanwhile, the Joliot-Curie family also took to the road. Irène was exhausted, so Frédéric left her and the children at a sanatorium in Salagnac in the Dordogne, and continued on his way. When he arrived in Bordeaux the next day, he struggled to make up his mind, eventually deciding to stay in France. He wanted to talk to Halban and Kowarski one last time, but was unable to locate the *SS Broompark*. The ship had departed several hours early on its way down the Gironde, in the hopes of avoiding interception by the Germans.[16] Joliot remained in France for the duration of the war and went on to play an important role in the Resistance.

The journey up the Atlantic coast and across the Channel took over 36 hours, as the steamship maneuvered to avoid the German U-boats. The two scientists and their families arrived in the town of Falmouth, in Cornwall, then took a train, along with their precious cargo, to London, in late June of 1940. Halban managed to convince the British of the importance of their research, and John Cockcroft, a British physicist and student of Rutherford some 10 years older than the two French refugees, came to their assistance, offering to set them up in the Cavendish Laboratory at the University of Cambridge.

To get from London to Cambridge, the two men borrowed a car and set out on a mad journey. Fearing a German invasion, the towns and villages between the two cities had removed every last signpost and, as the men were foreigners, they weren't allowed a roadmap. Kowarski decided the best course of action was to memorize the names of the pubs in each village along their route, and that was how they found their way![17] It wasn't until that August that they were able to resume their research, in the very laboratory where the neutron had been discovered. The two men knew there wasn't a second to lose: there was a scientific race underway…

The British Take the Lead

England, the country where Halban and Kowarski ended up after leaving France, had become a refuge for numerous other researchers, too. Since the late 1930s, some thirty renowned experts, many of them of Jewish heritage, had fled there to escape the wave of Nazi terror that was spreading across Europe. Understandably, they were highly motivated at the prospect of helping the Allies in their fight against the Axis powers. Several of them had experience in nuclear fission, the key physical phenomenon involved in building the bomb. Among them were Otto Frisch and Rudolf Peierls, who were on the verge of making an extraordinary and decisive discovery. In March 1940, the two Jewish physicists, who had sought refuge in Birmingham, calculated that the mass of uranium required to make an atomic bomb was actually much lower than originally thought: somewhere in the order of one kilogram. They wrote a brief, top-secret memo announcing the news and explaining the consequences of an atomic bomb, and sent it to Mark Oliphant, a British professor of physics, who passed it up the chain to the government. Frisch and Peierls' memo caused a considerable stir among the British authorities.[18] Until then, it had been believed the critical mass of uranium required to build a bomb was several tonnes, an assumption that rendered its construction purely hypothetical.[19]

In response to the memo, then-Prime Minister Winston Churchill created the MAUD committee and tasked it with examining the matter and subsequently presenting a report to the government. The name of the committee came about via a rather strange turn of events. When the Germans invaded Denmark, Niels Bohr, the most famous physicist in the world, after Einstein, sent a telegram to Otto Frisch, signing off with the words "Inform Cockcroft and Maud Ray Kent." No doubt caught up in the paranoid atmosphere of wartime, Otto Frisch, who knew no-one by the name of Maud Ray Kent, immediately set to work trying to crack the hidden code that surely referred to uranium disintegration ("ud" being the last two letters of the name "Maud"). Try as he might, Frisch was unable to decrypt the message, and eventually it was decided to use the word as the committee's codename. As it turns out, Bohr was simply sending along his best wishes to his children's former governess who lived in Kent, a woman by the name of Maud Ray!

The MAUD committee consisted of eight people, including Mark Oliphant, James Chadwick (the man who discovered the neutron), and John Cockcroft, a physicist at the Cavendish Laboratory, who would later end up in Montreal. These three men had previously worked with Ernest Rutherford at Cambridge. It was "all hands on deck" as the universities of Liverpool, Oxford, Cambridge, and Birmingham, as well as the Imperial Chemical Industries (ICI) company, were all called on to contribute. The MAUD committee held its first meeting on April 10, 1940. But if they were to build a bomb, they would need the right kind of uranium! Two types of uranium are found in nature—uranium-235 and uranium-238—and both are closely related. The only thing that distinguishes them is their slightly differing mass. Soon after the discovery of fission, scientists found that only uranium-235 would explode when it absorbed a neutron. If a bomb was to be built, it would mean separating the uranium-235, which forms only a fraction of naturally occurring uranium (less than 1 percent), from uranium-238. This process would prove to be very difficult, precisely because the two types of uranium are so similar, from a chemical standpoint. Practical and theoretical research programs into uranium separation (enrichment) were launched, primarily at the University of Liverpool and at the Cavendish Laboratory at the University of Cambridge, research that continued, along with the committee's work, for more than a year.

Halban was determined to remain in the thick of things, and wanted to be a fully fledged member of the MAUD committee, but he was informed that the official rules forbade foreigners from sitting on defence committees. It was Cockcroft who once again came to Halban's rescue, creating a technical sub-committee that the two Frenchmen were allowed to join. As a result, Halban was invited to virtually every meeting of the MAUD committee. Given the pace at which research was proceeding in Great Britain, in the spring of 1941, the committee was convinced it was feasible to produce enriched uranium (U-235) to build a bomb. The British were counting on Canada's large reserves of uranium to supply them during the war, and began talks with the government of the former colony to secure them. The Canadian government consequently purchased share control of the Eldorado Gold Mines Company, which operated a uranium mine near Great Bear Lake, in the Northwest Territories. The uranium ore was then transported nearly 4,000 kilometres south to Port Hope, on the shores of Lake Ontario, to be refined.

Once all the logistics were worked out, the MAUD committee approved the final version of two reports on July 15, 1941. The first discussed how uranium could be used to build a bomb, and was based on the research conducted by Peierls and Frisch, while the second touted uranium as a source of energy, the angle Halban and Kowarski were working at Cambridge. Both documents were submitted to the government. The report describing the new weapon provided technical details for producing an atomic bomb, estimating at 12 kilograms the critical mass of uranium-235 required to build a bomb equivalent to 1,800 tonnes of TNT. That was a whopping ten thousand times the power of the biggest conventional bombs at the time!

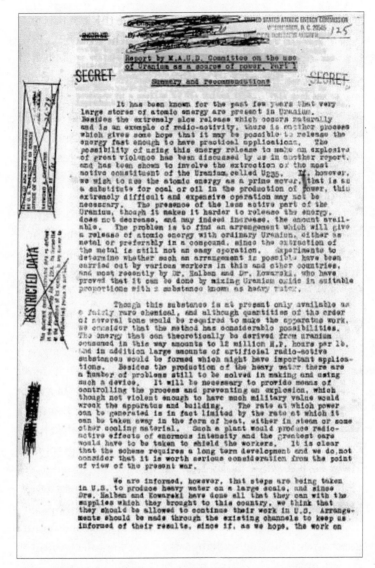

Report by M.A.U.D. Committee on the use of Uranium as a source of power. Part I

Summary and recommendations

SECRET   SECRET

It has been known for the past few years that very large stores of atomic energy are present in Uranium. Besides the extremely slow release which occurs naturally and is an example of radio-activity, there is another process which gives some hope that it may be possible to release the energy fast enough to have practical applications. The possibility of using this energy release as an explosive of great violence has been discussed by us in another report, and has been shown to involve the extraction of the most active constituent of the Uranium, called U235. If, however, we wish to use the atomic energy as a prime mover, that is as a substitute for coal or oil in the production of power, this extremely difficult and expensive operation may not be necessary. The presence of the less active part of the Uranium, though it makes it harder to release the energy, does not decrease, and may indeed increase, the amount available. The problem is to find an arrangement which will give a release of atomic energy with ordinary Uranium, either as metal or preferably in a compound, since the extraction of the metal is still not an easy operation. Experiments to determine whether such an arrangement is possible have been carried out by various workers in this and other countries, and most recently by Dr. Halban and Dr. Kowarski, who have proved that it can be done by mixing Uranium oxide in suitable proportions with a substance known as heavy water.

Though this substance is at present only available as a fairly rare chemical, and although quantities of the order of several tons would be required to make the apparatus work, we consider that the method has considerable possibilities. The energy that can theoretically be derived from Uranium consumed in this way amounts to 12 million H.P. hours per lb. and in addition large amounts of artificial radio-active substances would be formed which might have important applications. Besides the production of the heavy water there are a number of problems still to be solved in making and using such a device. It will be necessary to provide means of controlling the process and preventing an explosion, which though not violent enough to have much military value would wreck the apparatus and building. The rate at which power can be generated is in fact limited by the rate at which it can be taken away in the form of heat, either in steam or some other cooling material. Such a plant would produce radio-active effects of enormous intensity and the greatest care would have to be taken to shield the workers. It is clear that the scheme requires a long term development and we do not consider that it is worth serious consideration from the point of view of the present war.

We are informed, however, that steps are being taken in U.S. to produce heavy water on a large scale, and since Drs. Halban and Kowarski have done all that they can with the supplies which they brought to this country, we think that they should be allowed to continue their work in U.S. Arrangements should be made through the existing channels to keep us informed of their results, since if, as we hope, the work on

The MAUD committee report (first page) on the use of uranium as a source of power was the impetus for the work carried out at the Montreal Laboratory. (Government of Great Britain.)

The British immediately realized that this new weapon would be decisive in the outcome of the war, and were very concerned what would happen if the enemy got their hands on it. Germany had already managed to secure some of the essential resources, including the uranium mines of Joachimsthal when it invaded Bohemia in March 1939, as well as Belgium's uranium reserves in the Congo. What's more, when it occupied Norway in June 1940, it took over the Vemork heavy water production facility. The Nazis also had some heavyweight nuclear scientists, namely Otto Hahn, future winner of the Nobel Prize in Chemistry, and Werner Heisenberg, one of the pioneers of quantum mechanics. This meant the Germans had the material, financial, and human resources they needed to rival Great Britain in the increasingly frantic atomic race.

With the increase in the production of heavy water at Vemork, along with a number of other indicators, the Brits' suspicions were confirmed—the Nazis, too, were preparing to build a bomb. After several months of discussions at the top levels of the British government, in October 1941, Winston Churchill approved the plan to develop the bomb, and ordered the project to be given official status. ICI chemist Wallace Akers was appointed to head up the operations. To mask the actual purpose of the project, it was dubbed codename "Tube Alloys," suggesting, to the uninitiated, that it had something to do with the aeronautics industry.[20] The MAUD committee recommended at that time that the government consider collaborating on the project with the Americans.

England Under Threat

Throughout this period, from the invasion of France (May 1940) until the United States entered into the war (December 1941), England's situation was precarious, and the country feared it would be invaded. In the meantime, hundreds of German bombers pounded the island from the skies. London was heavily damaged, as were Plymouth, Birmingham, and Liverpool. Three and a half million Britons were evacuated, nearly 50,000 civilians were killed, and over 100,000 were wounded. The entire country was under siege. Ken Chackett, a chemist who would later work at the Montreal Laboratory, used to act as a night watchman at the top of the University of Birmingham clocktower, and when incendiary bombs were dropped on his lookout post, he would shove them over the edge of the tower, to prevent it from catching fire.[21] The avowed aim of the German operation was to demoralize the population and force the Churchill government to sign an armistice. But the Londoners held firm, and Churchill refused to sign any peace agreement with Hitler.

The British sought to turn the tide of the war in any way they could, and one such means was to seize the initiative on cutting-edge scientific projects. These included radar research conducted at Malvern College near Birmingham. Then, there was the research carried out at Bletchley Park, halfway between Oxford and Cambridge, where celebrated mathematician Alan Turing sought to break the secret codes the Germans encrypted using the Enigma machine. The atomic project was not yet a top priority, and Tube Alloys had to fight to recruit the scientists it needed.[22]

The groups hired by the MAUD committee continued their work nonetheless, as did Halban and Kowarski, who

conducted their research on the atom at the Cavendish
Laboratory. They were supported by new groups at the
universities of Bristol and Birmingham. In an attempt to
breathe new life into its programs, Britain turned to its
former colonies. Only too aware of the constraints they had
been under since September 1940, the British hoped to bring
the Americans on board, and thereby attract more funding
and resources. A British scientific delegation known as the
Tizard mission (so named after the mission head) was dis-
patched to the United States to establish scientific cooperation between the two countries. Its members included John
Cockcroft, one of Britain's leading atomic research scientists.
When Cockcroft met Enrico Fermi at Columbia University
in New York, he found to his amazement that Fermi was
pursuing the same research leads as Halban and Kowarski
at Cambridge.

The main difference between the two groups was the
type of nuclear reactor they were seeking to build. While the
Halban-Kowarski team was using heavy water as a moderator, the team headed by Fermi, an Italian physicist recently
arrived at Columbia, was using graphite. At that time, Fermi
and the American atomic researchers were several months
behind Halban in their research. An information-sharing
protocol was drawn up, and the British forwarded their main
findings, along with the two MAUD committee reports, to
the American Uranium Committee. The American group
was chaired by Lyman Briggs, a Roosevelt-appointed scientist who didn't believe in the feasibility of an atomic bomb.
As a result, Briggs simply sat on the information and locked
the reports away in his safe! In the summer of 1941, physicist
Mark Oliphant went back to the United States to push for
Anglo-American cooperation on the matter. When he dis-

covered that Briggs had failed to share the findings, Oliphant took it upon himself to inform the other members of the American Uranium Committee that the atomic bomb was in fact a real possibility.

A short time later, the Americans set up their own mission. Two renowned scientists, Harold Urey and George Pegram, travelled to England to visit the British laboratories working on the Tube Alloys project. Roosevelt even wrote to Churchill suggesting the two countries collaborate. The British, failing to realize that the Americans had begun to view the endeavor with the utmost seriousness, were in no hurry to follow up on the offer. They eventually came around, but by then it was too late... Urey and Pegram returned to the United States and submitted their report, prompting President Roosevelt to launch a research program on the atomic bomb by creating what was to become the Manhattan Project. The very next day, December 7, 1941, the Japanese attacked the Americans at Pearl Harbor, and the U.S. immediately entered into the world conflict and deployed their vast resources. A number of secret cities were erected, including Los Alamos in New Mexico and Oak Ridge in Tennessee, each of which was devoted to exploring different techniques for building an atomic bomb. Caught off-guard by this sudden surge in research investment, several British scientists urged the Churchill government to come to an agreement with the Americans so the two countries could collaborate on the project. When they realized they were now trailing the Americans on the research front, in the summer of 1942 the British requested that their Cavendish group be incorporated into the Chicago Laboratory. The Americans refused, confident they no longer needed Great Britain's help to develop the bomb, and convinced that the weapon would be of vital importance in the political reconfiguration that was bound to result once the war ended. So, they decided to limit the exchange of information with the British to the strict minimum.

And the situation only worsened once the U.S. Army took over the atomic project in June 1942, under the leadership of the Manhattan District's Army Corps of Engineers. In September 1942, responsibility for the Manhattan Project, as it was known, was transferred to Lieutenant General Leslie Groves, who was assisted by physicist Robert Oppenheimer. Groves immediately brought in draconian security measures. He feared espionage by the communists, and harbored a deep mistrust for the many foreign scientists working in Cambridge for the British, so he severed all scientific ties with the English.

Despite all these military and political twists, Halban and Kowarski's group continued their research in England. From the moment they arrived in the country, they picked up their experiments where they had left off with the Joliots in Paris. By immersing uranium in the heavy water they had carried with them from France, within the space of a few months, they were able to conclude, with near-absolute certainty, that it was possible to create a chain reaction. However, due to the shortage of experimental materials, they were unable to take their research to the next stage and actually achieve such a reaction. What's more, the low concentration of uranium-235 contained in the natural material prevented them—although they didn't realize it at the time—from obtaining the right fuel to make a bomb. So, they focused instead on the second application recommended by the MAUD committee: energy production. But in the context of the fierce and unrelenting war against the Germans, this avenue was not a priority, and it wasn't until a discovery was made by their neighbours that Tube Alloys' interest in developing a nuclear weapon was rekindled.

Based on their calculations, Egon Bretscher and Norman Feather, two physicists at the Cavendish Laboratory, suggested that, when uranium-238 absorbs a neutron, it is transformed, through radioactive decay, into plutonium-239. If their predictions were accurate, by absorbing a neutron, this new atom could split into two (fission) and release the energy needed for a bomb. At that stage it was still speculation, as the two scientists did not have the equipment to verify their claim, but it was a serious hypothesis that opened up a new avenue for building a bomb, this time using plutonium.[23] Although Halban's group didn't immediately realize it, the discovery nevertheless marked a major turning point for them. Now, there were two possible routes to an atomic bomb: one relying on enriched uranium-235, which involved considerable technical challenges, and the second using plutonium, which required building an atomic pile (reactor) inside which plutonium would be produced in sufficient quantities.

This second option was potentially more straightforward than the uranium bomb, as it appeared that plutonium was easier to isolate. The atomic pile envisaged by Halban and Kowarski, using natural uranium and heavy water, would enable them to manufacture plutonium. This avenue, which had been dismissed by the MAUD committee in 1941, subsequently became the clear choice for the British, boosting the profile of the Cavendish Laboratory in the process. It was against this backdrop that the decision was made to relocate Halban's group to North America where, it was hoped, the researchers would have access to the uranium and heavy water they needed to see the project through to fruition.

Montreal: The Project's New Home

In fact, for more than a year, there had been talk in London of transferring the project to North America. Halban's group was keen to join Fermi, who was working on a uranium pile in Chicago. Unfortunately for the British scientists, the American authorities didn't share their enthusiasm. So, they turned instead to a trusted ally with vast resources of its own: Canada. The former colony was decidedly less reticent than its neighbour to the south.[24] In Ottawa, the Minister of Munitions and Supply, C. D. Howe—nicknamed the "Minister of Everything" because he was involved in so many dossiers—was actually extremely favourable to the idea. An agreement was promptly signed between the two countries in the utmost secrecy: only the prime ministers and several senior members of the government were aware of its existence. It was agreed that Britain would pay the salaries of the British employees, and that the National Research Council of Canada (NRC) would cover the costs of the Canadian employees, as well as most of the costs of the research premises and equipment. Originally, the city of Ottawa was tapped to host the laboratory, but in the end, it was Montreal that got the nod, thanks to its modern airport, rail network, and two leading universities (McGill and Université de Montréal). It was also thought that the presence of a laboratory was more likely to go unnoticed in Montreal than in Ottawa, a capital city home to many international embassies.

All that remained was to appoint someone to head up the brand-new Montreal Laboratory. It was exactly the kind of position Hans Halban had been dreaming of since he arrived in England. And in the fall of 1942, he was appointed to lead the research group. Halban, on paper at least, seemed the obvious choice, but in actual fact his appointment had the effect of hampering the project's progress. The first signs of dissent came from Halban's closest collaborator, Lew Kowarski. The latter, whom Halban had always considered his subordinate, decided to remain in Cambridge to get out from under Halban's thumb. There was a sense of resentment between the two men. Kowarski was rather socially inept, and came from a family of little means, which immediately set him at odds with Halban, a man of refined upbringing. What's more, Kowarski had no patience for what he considered his colleague's political pandering. And in fact, for the duration of 1942, Kowarski headed up the research groups at Cambridge while Halban spent most of his time either travelling or in administrative meetings. Most of the Cambridge-based scientists threw their support behind Kowarski, refusing to be transferred to Montreal. It took all the diplomacy and prestige James Chadwick (discoverer of the neutron) and Edward Appleton (secretary of DSIR, the British equivalent of NRC) could muster to quell the mini-rebellion and have the group eventually relocated to Montreal in February 1943.

Atomic science was still in its fledgling stages, and specialists were few and far between, either already snapped up by the Americans or dispersed in hostile countries. Halban began by hiring Georg Placzek, a Czech national and renowned physicist from a well-to-do Jewish family, who had worked with Bohr on the uranium-235 problem. In fact, it was at Bohr's lab that the two physicists first met in the mid-1930s. Since the beginning of the war, Placzek had been teaching at Cornell University in New York State. He was hired by Tube Alloys and spent the fall of 1942 at

Cambridge with Kowarski, before arriving in Montreal in late December. Placzek spoke several languages fluently, making him a highly valued member of the cosmopolitan team. He was appointed to head up the group's theoretical physics division, and assisted Halban in recruiting researchers. The two men managed to attract some very high-level scientists who were at the leading edge of atomic physics. The only problem for the British and Canadian authorities was that they were all foreigners. Fritz Paneth (an Austrian) was hired to lead the chemistry group, while Pierre Auger (a Frenchman) was brought on board to head up the experimental physics group. Halban and Placzek travelled to the U.S. to interview several scientists who had sought refuge there and hadn't been employed to work on the Manhattan Project (the Americans didn't trust foreigners). The physicists who had worked with Fermi were of particular interest to the pair. One of them, Bruno Pontecorvo, who had been sent by the Americans to work for an oil company in Oklahoma, readily agreed to join the Montreal group.

But recruiting some of the other scientists was more of a challenge. The case of Franco Rasetti is a prime example of the difficulties Halban and Placzek encountered in trying to put together a topnotch team. Rasetti, an Italian physicist who had fled Mussolini's racial laws, was teaching at Université Laval in Quebec City. His profile was exactly what they were looking for: Rasetti had been Fermi's first and closest collaborator. Since arriving in North America, he had been conducting experiments on cosmic rays and had published six articles on his work in various scientific journals. Halban and Placzek reached out to him right away. Rasetti was interviewed by the pair at the Ritz-Carlton Hotel in Montreal. Placzek knew the Italian physicist well for hav-

Franco Rasetti preferred to continue teaching physics at Université Laval in Quebec City rather than join the Montreal Laboratory. (Paul Koenig, courtesy of Emilio Segrè Visual Archives, American Institute of Physics.)

ing worked with him in Rome in the early 1930s. The meeting went well, and Rasetti was especially excited to learn that another Italian, Pontecorvo, was slated to arrive in the city shortly. There was just one snag... his salary.

In a 1943 entry in his diary, made public for the first time in this book, Halban brings up the matter on more than one occasion:

> Thursday, February 25, 1943 — I informed Mackenzie [director of NRC] that Rasetti is in principle prepared to join us, but that he earns $6,000 at Laval University in Quebec, and that he would expect to get some more here in order to

indemnify him for moving and for the higher cost of living in Montreal. Bauer [a German and close engineer friend of Halban's and one of the first to arrive in Montreal], with whom I discussed this, thinks that there will be great administrative difficulties on this account, as the salary of Rasetti is already very high.

I suggested to Mackenzie that it might be possible to obtain Rasetti on loan from his University, the University paying the salary as a gift to the country. In this case the Research Council would only have to pay a small indemnification for Rasetti's living away from Quebec. Mackenzie thinks this plan very feasible.

[...]

Wednesday, March 31, 1943 — Saw Placzek and later had longer conversation with Auger, Placzek and Pontecorvo concerning Rasetti. They all three assure me that the rather individual character of Rasetti will not, to the best of their belief, disturb the working of the Laboratory.

[...]

Wednesday, April 7, 1943 — In the afternoon, great scene with Laurence [physicist in charge of recruiting Canadian scientists] who protested, in the presence of Placzek and Auger, in Auger's room, against the salary we intend to pay Rasetti, although he agreed to it at the Special Meeting of the Committee ... I think, however, that a weak point with Rasetti is that he ought to have accepted to come here for the same salary he has in Quebec.

[...]

Friday, April 9, 1943 — ... Mackenzie arrives and stays till the beginning of the meeting. I showed him the whole Lab, and I have a short conversation concerning Rasetti with him. He assures me once more that he is in full agreement that we take him on. He would, of course, be glad that we do not raise his salary too high. He is quite prepared to give us an introduction to the Archbishop of Ottawa in case we have difficulties about getting Rasetti released from Laval University.

An offer was finally sent to Rasetti on April 14, 1943, but the physicist turned it down, preferring instead to remain at Université Laval until 1947. In a discussion with Oliphant (a British physicist), Halban had this to say about it: "... and that I also tried to obtain Rasetti, even at the risk of grave injury to Canadian relations, and that he did not come only because at the last moment he demanded too big a salary." After the war, Rasetti would claim that he'd turned down the offer to work at the Montreal Laboratory for moral reasons (because of the plutonium being produced there to make a bomb), however, Halban's diary suggests that another, somewhat less noble, reason may have played an equally important role in his decision. It was, nevertheless, unfortunate for Halban and Placzek to have Rasetti slip through their fingers, as there were very few physicists of his calibre—and, especially, of his experience—at the Montreal Laboratory. Having Rasetti onboard would have breathed new life into the team.

The Americans had managed, very early on, to get their hands on Fermi and, as we shall see, the Italian scientist helped them take a giant step forward in the race to build a nuclear bomb. The Montreal Laboratory would have to look closer to home to find the resources it needed to complete its team.

Chaotic Beginnings

Given the situation, Hans Halban resigned himself to recruiting the people he needed in Montreal, even though his network of contacts there was rather limited. He turned to a local resource, physicist Pierre Demers, one of the only French-speaking Quebecers to work at the Laboratory. Demers was known by the French, as he had worked at the Joliot-Curie Laboratory in Paris. He was interviewed by Halban at the Mount Royal Hotel, and joined the British project in December 1942. The other Canadians on the team were recruited by George Laurence, one of the great Rutherford's former collaborators, and a man who had, in the 1930s, started up the only nuclear research group in Canada, as part of NRC. Laurence managed to recruit a good number of young physicists, mathematicians, and chemists who would make up nearly half of the employees at the Montreal Laboratory. He also hired several young women who were accomplished in mathematics as "computers." These included Joan Wilkie, Gilberte Leroux, and Fernande Rioux. They would spend their days performing mathematical calculations for the Lab's theoretical physics research. This was in the days before electronic computers, and women were considered to have greater concentration skills than men for this kind of task.

The first research conducted at the Simpson Street premises was indeed theoretical in nature, since the material required (heavy water and uranium) had yet to arrive in Montreal. The scientists published a number of studies in the early months of 1943, including the MT-4 Report[25] drafted by the Montreal Laboratory entitled "Notes on Diffusion of Neutrons Without Change in Energy." Physicist M. M. R. Williams, in a recent article, noted that this report was very important, as it laid the groundwork for neutronics, the science that studies the behaviour of neutrons in a nuclear reactor. In the beginning, the team's chemists had to content themselves with reviewing the literature, since they

Joan Wilkie was the first "computer" hired by the Montreal Laboratory, in 1943. This portrait of her was painted by Alma Duncan, a friend of the family. (Library and Archives Canada, Alma Duncan and Audrey McLaren fonds, R3209-1-6-E.)

had no material of their own to work with. Their first report dealt with the chemical effects of radiation, and presented a summary of the articles published on the topic between 1927 and 1942.

Fortunately, the first contingent of British scientists, consisting of 15 people (21, including their families), arrived by train from Halifax on the evening of January 24, while a second contingent, including Jules Guéron, a French chemist whose salary was paid by General de Gaulle's Free France, arrived on February 9. The long-awaited barrels of heavy water that had been spirited out of France by Halban and Kowarski, as well as the crates of equipment from Cambridge arrived two weeks later. At last, the scientists could get down to work, but they soon realized there wasn't enough space to house everyone at the Simpson Street villa. In his welcome speech to the newly arrived scientists, Halban announced that plans were in the works to relocate to a larger premises. In the meantime, they could visit the laboratory on Simpson Street, but only in small groups. Having arrived a couple of weeks too soon, they would have to make do with getting to know the other members of their division and reading reports. They were even given time off to find accommodation for themselves. The clock was ticking, and research was still at a virtual standstill.

In addition to the problems of procuring material and finding adequate premises, Halban had concerns of his own. Although few of his colleagues were aware of it, Halban was preoccupied by the fate of the patent application he had filed in May 1939. The process for which he sought a patent was entitled *"Perfectionnement aux charges explosives"* (Method for perfecting explosive charges) and it described the principle of an atomic bomb. The document had been co-authored by four scientists: Halban, Kowarski, Joliot-Curie, and Francis Perrin, another French physicist, and they had assigned the patent to Centre National de la Recherche Scientifique (CNRS, France's scientific research centre) before the Nazis arrived. Upon arriving in England, Halban fought to have the French patent recognized and its priority applied. He devoted considerable energy to the matter, as he felt it was his duty to protect France's interests, which Joliot-Curie had made clear to Halban before he fled to Bordeaux. After much negotiating, an agreement to share the royalties between the British researchers and their French counterparts was finally signed in July 1942, two years after they arrived in Cambridge. Even after arriving in Montreal, he continued to devote time to the legal aspects of his patents, as evidenced by his journal entry dated March 23, 1943: "3:30, Signed 7 patent applications at U.S. Consulate." The whole issue of patents was, in fact, to become a thorn in the side of the Montreal Laboratory in its relationship with the Manhattan Project.

There was something else bothering Halban, too. As soon as the Cambridge scientists arrived in Montreal, Halban questioned them about the "Kowarski Affair." He was keen to find out why his former colleague had refused to play second fiddle to him in Montreal, and sought to assess the degree to which the physicists supported him. He met, one-on-one or in groups, with all the specialists who had worked for Kowarski in England. In a telling conversation he had with Mark Oliphant in September 1943, it is clear what Halban thought about his former colleague and how preoccupied he was by the whole matter:

Tuesday, Sept. 21—Continuing with Oliphant, having a long and very pleasant conversation on the Kowarski problem [...]. I told him that after the way he behaved when he tried to make members of the Lab refuse to come over here, and the fact that he is not an independent worker and has great need of a boss to look after him, I do not see any possibility for successful collaboration with Kowarski. Oliphant tells me frankly that people like Chadwick and Compton think that I am against Kowarski because of competition, but I told him that he ought rather to look at the people I got, with big effort, to collaborate, such as Placzek, Pontecorvo, Auger, Laurence, who are all, even Laurence, exceedingly more competent than Kowarski […]. Oliphant asks how I think Kowarski would fit in as our ambassador in Chicago. I told him that our ambassador in Chicago should have my confidence, and that above all, he would not know how to cope with the feuds which are always suddenly developing in Chicago.

The chaotic beginnings of the Montreal Laboratory were unfortunate for the Tube Alloys project because, meanwhile, south of the border, it was full steam ahead. Fermi's group, in particular, managed to solve any problems that arose, in part due to the fact that their project was under the authority of the U.S. Army. Fermi had opted for a different moderator than Halban and Kowarski: graphite, as opposed to heavy water. He set to work building increasingly bigger piles at Columbia University's Department of Physics in New York City. In fact, they were so tall that there was no room big enough at the faculty for the pile to reach the critical phase required to trigger a self-sustaining chain reaction. So, the army decided to relocate the entire team to the University of Chicago, which had a large space (the school's former squash courts) beneath the stands of the varsity football stadium. And that was where, on December 2, 1942, the world's first nuclear reactor achieved criticality. The momentous event occurred in the late afternoon, after Fermi gave the order to remove the final cadmium rod from the core of the reactor.

As Laura Fermi described in her biography of her husband, that evening they threw a big reception at their home in Chicago. On the army's orders, the spouses were kept in the dark about the work their husbands were conducting, and were unaware of the true objectives of their research. To Laura Fermi's great surprise, all the guests who arrived at their home that evening rushed to extend their warm congratulations to her husband. Champagne was produced and the mood was celebratory. Try as she might, Laura was unable to learn what had caused such excitement among the physicists. And she didn't discover the truth until August 1945…

In a twist of irony, Hans Halban happened to be in Chicago at the time of that technical triumph that marked a decisive victory for the Americans. But that day, December 2, 1942, Halban wasn't by Fermi's side; he was on a recruiting mission in the city to try and attract talent to continue the race with that country's neighbours. One of the candidates he met with *was* on hand to witness the Chicago Laboratory's great success. His name was Bertrand Goldschmidt, and he would go on to play a leading role in the race to build the atomic bomb.

The Laboratory relocated to the new pavilion of the Université de Montréal in 1943 situated on the northwest side of Mount Royal. (Aerial Photo, 1945, Canadian Pacific Airlines, BAnQ 4751155_1)

LABORATORY AT WAR

Bertrand Goldschmidt, A Valuable Find

That day, December 2, 1942, the telephone rang at the National Defense Research Committee office in Washington. James Conant, chair of the committee that oversaw the Manhattan Project, picked up the receiver and listened in astonishment to the coded message that would go down in history: "You'll be interested to know that the Italian navigator has just landed in the New World." Read: Enrico Fermi has just triggered the first-ever nuclear chain reaction. Despite falling behind the British in the beginning, the Americans now had a significant lead in the atomic race. As it happened, the confidential message, which was destined for those within the American nuclear project only, was overheard by a young Frenchman who was there that day: Bertrand Goldschmidt. It was a series of highly improbable circumstances that led to Goldschmidt's chance discovery, one that would benefit, against all odds, the British (soon to be Montreal-based) atomic project.

And yet, things had started out badly on American soil for Goldschmidt, a French Jew who'd been forced to flee Europe to escape the Nazis. Once the relief of having eluded the German barbarity subsided, the French nuclear specialist realized that the United States was a gilded cage, as he was repeatedly denied access to the country's scientific labs throughout 1941. Even Fermi was unable to overcome the U.S. Army's mistrust of foreigners to hire the man who had been trained by the Curies at the Radium Institute. Frustrated at not being able to apply his crucial skills in the race to build the bomb, Goldschmidt eventually agreed, in January 1942, to volunteer his time at New York's Memorial Hospital. He could hardly turn down the mission, since it consisted of taking readings of cancer patients who were testing a radioactive phosphorous treatment. For Bertrand Goldschmidt, his volunteering turned into a blessing when he crossed paths with a man willing to go to great lengths to recruit scientists of his calibre. That man was none other than Hans Halban, who had been combing the Eastern Seaboard in the hopes of getting his hands on several experts in nuclear physics. The two men already knew each other, having met at the Curies' institute. After admonishing Goldschmidt for failing to have looked up his old French contacts, Halban promised to bring him on board his English team. A few months later, it was official: the British Department of Scientific and Industrial Research (DSIR) agreed to hire the young chemist. Hans Halban, aware that Fermi could very well beat him to the line by triggering the first chain reaction, took a gamble. He asked the Americans to take on Goldschmidt as an intern for a few months at the metallurgical lab in Chicago where the Italian genius worked.

Paradoxically, the Manhattan Project team no longer considered Bertrand Goldschmidt as a Frenchman, but rather as a member of a British laboratory, agreeing to take on the chemist in Chicago in the summer of 1942. Halban's gamble had paid off! One of his men was now firmly ensconced within the rival team. Like the Montreal Lab, the Chicago Lab's goal was two-fold: Enrico Fermi was to build a graphite-uranium atomic pile, while Glenn Seaborg was tasked with developing a plutonium extraction method. As a chemist, Goldschmidt joined the second group, and he was immediately apprised of the extraordinary nature of his task: He was to familiarize himself with the highly complex chemistry of a new element that did not exist in nature and that could be an extremely powerful explosive. Goldschmidt hadn't yet heard of plutonium. The Americans were producing it in particle accelerators by bombarding uranium. The amount of plutonium obtained this way was miniscule, and the extraction process was both difficult and risky.

This was the task Glenn Seaborg set for Goldschmidt. Extracting the plutonium meant first reducing the uranyl nitrate by dissolving it in ether, a highly volatile liquid that can explode in the presence of a flame. The chemists handling these solutions had to wear gloves and be protected behind lead screens. In his journal,[1] Seaborg recalls how Goldschmidt often forgot to wear rubber gloves and expressed surprise that a chemist with such experience with radioactivity could be so careless. Goldschmidt, in his book *Atomic Rivals: A Candid Memoir of Rivalries Among the Allies Over the Bomb*, comments: "Little did he know how few precautions we usually took at the Curie laboratory." Nonetheless, in the end, he managed to extract approximately 60 micrograms of plutonium. That's about thirty times less than the weight of a single bird feather!

But his experience would be worth its weight in gold in Montreal. Halban and Kowarski hoped to develop a nuclear pile like the one Fermi had built, using heavy water instead of graphite. They knew that once the reaction was launched, plutonium would be created, and this would be the perfect shortcut to the bomb, as it would avoid having to get bogged down with uranium-235 extraction. At that time, the number of chemists with expertise in plutonium could be counted on one hand, so the experience Goldschmidt acquired in Chicago was a real boon for the British project in Montreal, where he was put in charge of the plutonium section in the chemistry division headed by Fritz Paneth.

Bertrand Goldschmidt would make another significant contribution to the Montreal Laboratory: He would find the lab a site to finally conduct real experiments, a vast premises worthy of the team that was being assembled. For a clearer picture of how this opportunity came about, it's worth delving a little deeper into Bertrand Goldschmidt's past, to a time long before he fled to the United States. To 1939, to be precise, when the sound of gunfire was growing ever closer to France's borders…

Just before he was about to head off on a trip to Tahiti, Bertrand Goldschmidt attended a lecture by a man named Pierre Auger, a specialist in cosmic rays. Goldschmidt went up to speak to the scientist after his talk, and offered to take some cosmic ray readings at various northerly and southerly latitudes for him during his voyage. Auger agreed on the spot, and provided Goldschmidt with one of his detection devices as well as a brief training session at his laboratory at École normale supérieure. The French national scientific research centre (CNRS; Centre National de la Recherche Scientifique), headed by Henri Laugier, even freed up some funding so one of Auger's assistants could accompany Goldschmidt. The readings they took confirmed Auger's theory: The Earth's magnetic field has an effect on cosmic rays.

Goldschmidt's voyage home was eventful, as war had broken out while he was away. When his ship docked in Guadeloupe and Martinique, some of the passengers and crew were taken away in handcuffs and thrown into jail, the former for smuggling opium aboard and the latter for refusing to head back to France because the country was at war. The delay caused by these two incidents likely saved the lives of the travellers, as the ship was ultimately too late to join the convoy it had been assigned to for the voyage across the Atlantic, a crossing that saw one of the convoy's passenger liners torpedoed and sunk by a U-boat. The ship Goldschmidt was on zigzagged its way across the ocean in an attempt to confound the German submarines. It eventually made its way to Casablanca, before setting out again and docking in Marseilles on November 2, 1939.

After arriving in Canada more than three years later, Goldschmidt learned that Henri Laugier, the former director of CNRS whom he had met during the cosmic ray experiments, was now a professor at Université de Montréal. Goldschmidt happened to bump into him in mid-November at the airport in Montreal when he was on his way to visit his mother in New York City. In conversation, Henri Laugier mentioned he'd heard Halban was having trouble locating an appropriate space for his lab in Montreal. Goldschmidt was astounded that the news had leaked out. Laugier suggested he contact Université de Montréal since the university had a new building on Mount Royal that stood virtually empty. And so, in March 1943, the Canadian government rented out two floors in the west wing of the university's new pavilion for the Tube Alloys project. Physicist Pierre Auger would also get to know the new premises, as Goldschmidt suggested to Halban that he recruit him to head up the experimental physics division of the Montreal Laboratory. The cosmic ray experiment had served Montreal's scientific endeavour well!

Relocating to Université de Montréal

On January 8, 1943, Halban and Goldschmidt woke to an unpleasant surprise: the Laboratory was plastered across the morning paper's headlines. The front page of *Montréal-Matin* blared: "60 foreign scholars set up shop at Université de Montréal to pursue highly important research" (our translation). The article was accompanied by a photo of the new premises, which the team had yet to move into. The story went on to describe the researchers as "mostly of Israelite, Russian, French, Polish, Czech, and even German origin," explaining that they were at the university to carry out extensive research of the utmost importance on radioactivity, physics, and physical chemistry, under the leadership of a great French physicist, Mr. Auger, and the distinguished patronage of the National Research Council. So much for keeping a lid on their research into the atomic bomb!

Hans Halban was furious, especially since he was the one tasked with keeping the Laboratory's research secret. Leslie Groves, the lieutenant general in charge of the Manhattan Project, who was already more than a little paranoid about security, was unlikely to look kindly on this media snafu. What little trust the Americans had in the British project could very easily evaporate, causing Halban's team considerable harm. After a few phone calls, Halban found the culprit: it was Henri Laugier, the man behind the Laboratory's relocation to the Université de Montréal campus, who had spilled the beans.

Apparently, Laugier hadn't appreciated the lack of recognition for his key role in finding the new premises. He was reprimanded for his actions and, to avoid further leaks, the Canadian censors were informed. One week later, Halban, accompanied by Pierre Auger, went to meet the rector of Université de Montréal, Monseigneur Roux, to discuss the incident. The discussion between an atheist refugee of Jewish descent and the oh-so-Catholic university rector was a delicate affair. As Halban noted in his diary: "This morning, with Auger, I saw Monseigneur le Recteur Roux, and thanked him for the hospitality of the University. I told him that this was a combined U.K.-Canadian effort, and Monseigneur said that he was very upset about our having been embarrassed by journalistic activity. I explained to him the position concerning Dr Laugier's interference with 'La Presse' [Note from the author: Halban appears to have confused *Montréal-Matin* with *La Presse,* another Montreal newspaper]. The conversation was pronounced harmless and agreeable after that and Auger and I decided in the end, in order to give some sort of information, to say that we are doing radiological research, telling the Rector this fact was extremely secret."

Université de Montréal had been tight for space at its campus in the Latin Quarter since the late 1920s, and a new building had just been erected for it on the slopes of Mount Royal. Its Art Deco style was reminiscent of Quebec's many convents. But behind the symmetry of its façade was a warren-like maze of staircases and dead-end corridors. Former employee Annette Wolff described the building interior as the work of a crazy architect! To make matters worse, when the Tube Alloys project moved in, construction was still underway: running water was only available in parts of the building, and there were still a host of unresolved problems.

Nevertheless, Université de Montréal provided Hans Halban and his now-plentiful team with one major advantage—an abundance of space!

Finally, on March 27, several weeks after the initially scheduled start-up date, NRC took possession of its new premises in the university's west wing, and immediately moved in all the equipment, heavy water, and uranium that had been stored at the Simpson Street villa. It was Hans Halban himself who supervised transportation of the heavy water, a task he deemed vital. In the words of George Laurence: "This was a precious treasure. Great care was taken to avoid loss of even minute quantities by evaporation. Drips were carefully wiped up with hospital cotton which was then sealed in vapour-tight containers so that the heavy water could be recovered."[2] At last, the scientists, who had arrived in mid-February, had daily access to the Laboratory and sufficient space to begin their work. Two Royal Canadian Mounted Police officers were posted at the Laboratory entrance, and anyone wishing to enter had to show their credentials. Although the Laboratory employees were allowed to eat at the university cafeteria, they were instructed not to mingle with the students.[3]

Halban set the researchers two main goals: 1) develop the first atomic pile (or nuclear reactor, as it is known nowadays) outside the United States, and 2) study the chemistry of plutonium. There was a sense of great enthusiasm as the work got underway. While the theorists working under Georg Placzek calculated the properties of the future reactor and the amount of heavy water and uranium that would be needed, the chemists in Paneth and Guéron's division got to work measuring the impurities in the uranium, an essential piece of the puzzle to determine whether neutrons would be absorbed in large quantities by these impurities. However, the enthusiasm soon waned, as the supply of materials the scientists required for their research dried up, in large part due to the Americans' decision to deny access to them.

The Montreal Laboratory still had the heavy water—approximately 180 kilograms of it—that had been rescued by Kowarski and Halban in 1940. Based on the experiments conducted at Cambridge in 1941 and 1942, the theorists estimated that somewhere between three and six tonnes of heavy water were needed to build a decent-size reactor. That meant locating a source of heavy water so they could continue their work. As it turned out, there was a plant similar to the one in Vemork, Norway, in operation in Canada, in the town of Trail, British Columbia. The Cominco (Consolidated Mining and Smelting Company of Canada) plant produced ammonia for use in manufacturing fertilizer, and was able to add the production of heavy water to its process. But the Manhattan Project was looking for the raw material, too, and the Americans had beaten the Montreal group to it.

In 1942, Leslie Groves had signed a contract with Cominco, which agreed to provide heavy water to the Americans, at cost, for the duration of the war. In exchange, an investment of $2.8 million (an enormous sum at the time) was made to build the facility, which came into commission in June 1943. Although the initial production of heavy water was limited (7 kg in June 1943), it quickly picked up, until the plant was producing 150 kg per month in 1944 and nearly 500 kg per month by 1945. When they learned of the exclusive contract between Cominco and the Americans, Halban and the British were devastated.

The same story played out for uranium. One of the world's biggest mines was located in Canada, at Great Bear

Lake, in the Northwest Territories. It belonged to Eldorado Gold Mines, a company in which the federal government, as mentioned earlier, was a minority shareholder. The British delegated the Canadian government to negotiate with the company to acquire its production. The minister in charge, C. D. Howe, put his blind trust in company president, Gilbert LaBine. Alas, LaBine was more concerned with the company's bottom line, and he turned around and signed a contract with a more profitable customer—the Americans—who secured a deal to buy Eldorado Mine's uranium production for the next three years. Once again, the Montreal Laboratory was left empty-handed.[4] In the end, the Canadian government's involvement hadn't carried much weight against the deep pockets of the Americans.

The desperate shortage of raw material made the researchers' work a huge challenge. To keep the experiments going, Bertrand Goldschmidt had an idea. Instead of heavy water and uranium, the teams could work instead on plutonium. In February 1943, realizing that the Americans were hoarding the resources, he headed to Chicago, with Halban's blessing, to try his luck. Goldschmidt had no problem getting into Seaborg's lab, because his pass hadn't been revoked when he'd left. He was welcomed warmly, and returned to Montreal not only with valuable information but also with two test tubes in his pocket, one filled with a fraction of the fission products he had separated and analyzed, the other with a few drops of liquid containing four micrograms of plutonium[5]—all right under the nose of General Groves, who would never have approved such a "loan." Goldschmidt's team at that time consisted of only himself and Frank Morgan, a young British chemist whom we will come back to later. They used the miniscule quantity of plutonium he had acquired to repeat the experiments carried out in Chicago and confirm the chemical properties of this new element. However, they didn't have enough to test extraction methods. Goldschmidt had saved the lab's skin with that tiny amount of product, but if they wanted to stay in the atomic race, they would need much more. If they weren't able to restore ties with the Americans, Montreal's nuclear adventure could well end right then and there.

The Quebec Conference

The status of the atomic bomb had changed since the beginning of the war. The British experimental project had been overtaken by the Americans. And now, this new weapon could well play a decisive role in the outcome of the war, meaning it had become a topic of discussion for the most senior advisors of Churchill and Roosevelt. The timing was right, because the British prime minister and the U.S. president had been meeting regularly to coordinate their efforts since the United States had joined the war, and the next summit was slated to be held in Quebec City. In early August of 1943, the Canadian government requisitioned the Château Frontenac to house the British and American delegations who would be attending the summit conference. Hotel Clarendon was also enlisted to accommodate the 150 or so journalists who would be covering the event, as well as the Canadian Censorship Branch, which vetted every press release before it could go to print. Winston Churchill arrived in Halifax aboard the *Queen Mary*, then he took a train to Charny, on the south shore of the St. Lawrence River. He crossed the Quebec Bridge by car on August 10, and settled in, along with his wife and youngest daughter, on the main floor of the Citadelle of Quebec, the second official residence

At the Quebec Conference in August 1943, Winston Churchill and Franklin D. Roosevelt signed a secret agreement on nuclear cooperation. Canada's prime minister William Lyon Mackenzie King (at left) was invited to attend only the ceremonial events. (William George Horton, Official War Photographer, Government of the United Kingdom.)

of Canada's Governor General. That was also where Roosevelt and his wife were housed for the duration of the conference, from August 17 to 24.[6]

Churchill proposed that Canada's prime minister Mackenzie King be present at all the meetings, but his suggestion was vetoed by the Americans, who feared it would create a precedent for subsequent meetings. In the end, most of the meetings Mackenzie King attended were merely ceremonial in nature. The goal of the conference was to plan military strategy against the Germans and Japan for the coming months. The decision was made to step up the Allied bombings of Germany and to focus on the attacks against Italy to obtain its unconditional surrender. Plans were also laid for the Normandy landing.

During the conference, a secret agreement—the *Agreement Governing Collaboration between the Authorities of the U.S.A and the U.K. in the matter of Tube Alloys*—was signed to coordinate the development of atomic research. The agreement saw the British project partially incorporated into the Manhattan Project. The British agreed to make several of their scientists available to the Americans. In exchange, they would receive copies of all the U.S. reports from the Manhattan Project as well as a promise of heavy water and uranium for the Montreal Laboratory. The British were so eager to resume their exchange of information with the Americans that four of their key physicists (Chadwick, Peierls, Simon, and Oliphant) flew to Washington the same day the agreement was signed, on August 19, 1943.

Despite the provisions of the Agreement, the Americans were slow to collaborate, mainly due to the reticence of General Groves, who disapproved of the motley band of scholars who hailed from countries across Europe. He was particularly suspicious of the French, whom he suspected of leaking information and of indiscretions that threatened the security of the project. He was also concerned that hidden among the "foreigners," i.e., those with nationality other than American, British, or Canadian, were communists who were secretly passing along information to the Soviet Union. As we will see in due course, he was not entirely wrong, however, his prejudices severely hampered his ability to identify the real culprits, in particular those who were leaking information to the U.S.S.R. In September 1943, shortly after the Quebec Conference wrapped up, Niels Bohr, one of the world's most highly respected physicists, fled from Denmark along with his son before the Nazis could get their hands on him. The British invited him to join the Tube Alloys project and, from there, Bohr was hand-picked to be part of the contingent that was sent to the Los Alamos Laboratory, the focal point for American nuclear research. If they weren't going to be able to build their own bomb before the end of the war, the British hoped to at least advance their knowledge to the point where they could launch their nuclear program as soon as possible once the war had ended.

The Americans, at the urging of General Groves, placed an additional condition on the Anglo-American-Canadian collaboration: They demanded that Hans Halban be replaced by a British citizen at the head of the Montreal Laboratory. Halban was not looked on kindly by the Americans. For one, they were furious that Halban had filed patent applications in England (notably, together with Kowarski), and were especially incensed at his attempts to file further applications in the United States. They considered his actions harmful and felt they hampered the transmission of confidential information from the U.S. project. In addition, they believed that if they were going to incorporate Tube Alloys, at least partially, into the Manhattan Project, then it should be a British citizen heading up the Montreal Laboratory, not a foreigner inclined, in their opinion, to put France's interests first.

Niels Bohr was the greatest physicist of the 20th century, after Albert Einstein. One of the pioneers of quantum mechanics, he founded an institute in Copenhagen that produced a host of Nobel Prize winners. Bohr was hired by the Tube Alloys project and spent time in Montreal. (Paul Ehrenfest, courtesy of Emilio Segrè Visual Archives, Ehrenfest Collection, American Institute of Physics.)

It wasn't only the Americans who wanted Halban gone; even some of his own team members were loath to support him. Several of the employees at the Montreal Laboratory even flatly accused him of impeding its scientific activities. His perceived haughtiness, extravagant lifestyle, and quickness to criticize deepened the divide between him and the other scientists on his team. Halban's diary entries give more insight into his views. For instance, he repeatedly criticizes Canadian scientist George Laurence, whom he considered ineffective. He insinuates that Laurence lacked diligence in recruiting Canadian scientists, and that he didn't understand the rudiments of neutron physics. Yet, Laurence was, together with Pierre Demers, one of the only Canadian scientists with experience in atomic physics prior to the war, having worked at the Cavendish Laboratory in the 1930s under Ernest Rutherford.

What's more, the letters on file at the National Archives of Canada contradict Halban's words. Laurence worked tirelessly from late 1942 through the first months of 1943, sending out more than one hundred letters in an attempt to recruit Canadian scientists for the Montreal Laboratory.[7] Several of his "catches" would go on to have important careers in the field, including George Volkoff, Jeanne LeCaine-Agnew (see sidebar, p. 77), and Harry Thode. In all, he recruited over twenty young chemists, physicists, and mathematicians in the early part of 1943. Even much later, the animosity between Laurence and Halban still lingered, as evidenced by a talk Laurence gave to the Canadian Association of Physicists, in which he said, "Von Halban was not accustomed to Canadian business methods. The administrative staff in Ottawa was equally unaccustomed to the Austrian lesser nobility. They spoke of the Montreal Laboratory as a madhouse—not an ordinary madhouse, but one that was run by the inmates."[8]

After the Quebec Conference, Halban lobbied the authorities, telephoned to Ottawa, telegraphed to London, and had meetings in Washington. But his efforts were in vain: he was losing his grip on the Laboratory. His personal life, too, was in disarray. When he first arrived in Montreal, Halban had been married, for nine years, to an upper-class Dutch woman, Els Andriesse, and they had a three-year-old daughter, Mauld Halban. The young family had survived a very stressful time when France was invaded by the Germans and they were forced to flee. The couple managed to hold together through that tumultuous time, but when they arrived in Montreal, for reasons that remain unclear, the Halbans separated. Although they were discreet about it, it was a small community, and word soon got out that the Halbans were living apart. In a surprising twist, in the fall of 1943, Georg Placzek, who was in many ways Halban's right-hand man at the Laboratory, announced that he and Els Andriesse were engaged to be married. The situation appears to have been accepted without acrimony. And Bertrand Goldschmidt, who found the whole thing rather amusing, even invented a term for the new relationship linking Halban and Placzek, referring to them as "husbands-in-law!"

To help take his mind off things, Goldschmidt, who was never at a loss for ideas, invited Halban to spend a weekend skiing in the Laurentians with a friend of his who was coming from New York for the occasion. That's how Halban met Aline de Gunzburg, a young, very well-off widow (her family owned the Ritz in Paris, and several oil refineries in Europe, among other holdings). The two immediately hit it off, and shortly after, they wed in New York City. Aline moved with her young son, Michel, to Montreal, where she lived in a house Halban had rented at 1297 Redpath Crescent, at the foot of Mount Royal. It was a house that later became famous in the history of Quebec as the home of James Cross, the diplomat who was kidnapped by the Front de libération du Québec on October 5, 1970, triggering what would become known as the October Crisis. Hans Halban's turbulent life didn't go over well with his superiors and his team. And by then, nothing could stop the Americans and save Private Halban…

Els Andriesse separated from Hans Halban and, in 1943, she married physicist Georg Placzek, Halban's right-hand man at the Montreal Laboratory. The couple are seen here, likely at the front entrance of the Ritz-Carlton. (Photographer unknown. A. Gottvald, M. Shifman, *George* Placzek: *A Nuclear Physicist's Odyssey*, Singapore: World Scientific Publishing Co., 2018, p. 44. Aleš Gottvald, personal archives of Michael Fuhrmann.)

PIERRE DEMERS

Pierre Demers was one of the only French-Canadian scientists who worked at the Montreal Laboratory. He was a colourful character who was strongly committed to various causes, including Quebec independence and the use of the French language in the world of science.

His parents had long dreamed of living in Europe, and Demers spent part of his childhood in Paris, then in the south of France, near Cannes, where he attended elementary school. The Demers family returned to Montreal in 1925. Pierre completed his classical education under the Jesuits in 1933 at Collège Sainte-Marie and Collège Jean-de-Brébeuf, where he developed a keen interest in science. He enrolled at Université de Montréal where, during the 1930s, he earned diplomas in physics and mathematics, as well as a master's degree in chemistry under the supervision of Léon Lortie. In 1938, Demers was awarded a scholarship from the Government of Quebec to study in Paris at École normale supérieure—one of the most prestigious research universities in France—where Pierre Auger taught. He went on to join the atomic laboratory at Collège de France. There, he worked under Joliot and Hans Halban, and met Lew Kowarski. Demers began work on a technique to detect elementary particles using photographic methods, which would come in very handy in Montreal. In June 1940, like the rest of Joliot's team, Demers was forced to leave Paris ahead of the German invasion. He headed south, travelled across Spain and, like thousands of others fleeing the Nazis, boarded a ship in Lisbon headed for New York City.

When he returned to Quebec, Pierre Demers was hired by Canadian Industries Limited (CIL)—the parent company of DIL mentioned earlier—in the company's R&D laboratory at its explosives plant in Beloeil, southeast of Montreal. In late 1942, he was recruited by Hans Halban to join the Montreal Laboratory, where he worked for the remainder of the war with the experimental physics group, first under the direction of Pierre Auger, and then under Alan Nunn May. He authored eleven reports while at the Laboratory, in the following fields: sources of polonium-beryllium neutrons and radium-beryllium neutrons, neutron detection using photographic emulsions, and the neptunium decay chain. Demers convinced several members of the Laboratory to give talks at the Montreal Physics and Chemistry Society, and between November 1944 and March 1945, six of the Lab's scientists spoke about topics ranging from meteorites (Fritz Paneth) to the chemistry of radioactive products (Bertrand Goldschmidt).

In his work at the Montreal Laboratory, Pierre Demers co-authored a major report on the neptunium decay chain. This was an important discovery because only four radioactive decay chains existed, the three others being uranium-238, uranium-235, and thorium-232. This research into neptunium was noteworthy among the work conducted at the Montreal Laboratory as it had no direct practical application.

After the war, Pierre Demers was employed as a physics professor at Université de Montréal while he was finishing up his PhD in physics in Paris. One of the members of his thesis examination committee was Pierre Auger. Pierre Demers taught physics from 1947 until his retirement in 1980.

Demers continued to use the particle detection technique he had learned at Collège de France in the 1930s to study cosmic rays and solar radiation. A rather eccentric man, Pierre Demers subsequently turned his attention to two topics that were not especially "fashionable" in physics departments in the second half of the 20th century: colour and the periodic table. It wasn't so much the subjects themselves that were original, but rather the manner in which he approached them.

Along with a number of college teachers and artists, including painter Alfred Pellan, he founded Centre québécois de la couleur, which sought to "create ties among the scientific, artistic, and industrial communities around the theme of colour."

In the 1990s, Demers proposed a new arrangement of the periodic table, one that, instead of being organized in parallel columns and lines in the classic layout, would be organized as a spiral. In his table, the maximum number of elements would be

September 1945 photo of the Montreal Laboratory's experimental physics group that includes Bruno Pontecorvo, Alan Nunn May, Pierre Demers, and Lew Kowarski. *Left to right, front row*: Anne Barbara Underhill, Sara Courant, E. Marks, Jean Clark, Mary Dysart, Jill Organ; *second row*: Gertrude Soleman, Harriet Anderson, Bernice Sargent, Alan Nunn May, Bruno Pontecorvo, Yvette Diamond, Freda Kinsey; *third row*: Ernest Courant, John Jelley, Dudley Borrow, Ted Hincks, Lloyd George Elliott, John Vernon Dunworth, Gordon Graham, David West, John Harvey; *fourth row*: Solly Gabriel Cohen, Patrick Cavanagh, Pierre Demers, Ron Maskell, Heinz Paneth, Neil Niemi, John Bayly, Denys Booker, Brian Flowers, Alan Munn, Louis Nirenberg, Lew Kowarski, Ted Cranshaw, George Volkoff. (Photographer unknown, National Research Council Canada, personal archives of Irène Kowarski.)

120. Professor Demers also proposed, at around the same time, that element 118 be named québécium, "in honour of his home-land." In November 2016, this element was named "oganesson" after Yuri Oganessian, director of the laboratory in Dubna, Russia, where a number of super-heavy chemical elements

Periodic table of the elements

	Alkali metals	Halogens
	Alkaline-earth metals	Noble gases
	Transition metals	Rare-earth elements (21, 39, 57–71) and lanthanoid elements (57–71 only)
	Other metals	
	Other nonmetals	Actinoid elements

period	group 1*	2	3	4	5	6	7	8	9	10	11	12	13	14	15	16	17	18
1	1 H																	2 He
2	3 Li	4 Be											5 B	6 C	7 N	8 O	9 F	10 Ne
3	11 Na	12 Mg											13 Al	14 Si	15 P	16 S	17 Cl	18 Ar
4	19 K	20 Ca	21 Sc	22 Ti	23 V	24 Cr	25 Mn	26 Fe	27 Co	28 Ni	29 Cu	30 Zn	31 Ga	32 Ge	33 As	34 Se	35 Br	36 Kr
5	37 Rb	38 Sr	39 Y	40 Zr	41 Nb	42 Mo	43 Tc	44 Ru	45 Rh	46 Pd	47 Ag	48 Cd	49 In	50 Sn	51 Sb	52 Te	53 I	54 Xe
6	55 Cs	56 Ba	57 La	72 Hf	73 Ta	74 W	75 Re	76 Os	77 Ir	78 Pt	79 Au	80 Hg	81 Tl	82 Pb	83 Bi	84 Po	85 At	86 Rn
7	87 Fr	88 Ra	89 Ac	104 Rf	105 Db	106 Sg	107 Bh	108 Hs	109 Mt	110 Ds	111 Rg	112 Cn	113 Nh	114 Fl	115 Mc	116 Lv	117 Ts	118 Og

lanthanoid series 6	58 Ce	59 Pr	60 Nd	61 Pm	62 Sm	63 Eu	64 Gd	65 Tb	66 Dy	67 Ho	68 Er	69 Tm	70 Yb	71 Lu
actinoid series 7	90 Th	91 Pa	92 U	93 Np	94 Pu	95 Am	96 Cm	97 Bk	98 Cf	99 Es	100 Fm	101 Md	102 No	103 Lr

*Numbering system adopted by the International Union of Pure and Applied Chemistry (IUPAC).   © Encyclopædia Britannica, Inc

Periodic Table of the Elements. Element 118, known since December 2015 as "oganesson" should have been called "québécium," according to Pierre Demers. (Wikipedia Commons.)

were discovered. As is the custom, the team that synthesizes a new element gets to choose the name for the periodic table.

Pierre Demers' idea of a new layout for the periodic table was an interesting one, but like many of his initiatives, it failed to garner widespread support.

In the social sphere, Pierre Demers earned a name for himself as an ardent defender of the French language, in particular its use in the scientific world. He never accepted the idea that English should be the universal language of the sciences. In 1980, he founded, along with nine others, including sculptor Armand Vaillancourt, Ligue internationale des scientifiques pour l'usage de la langue française (LISULF), an international league that promotes the use of French in the sciences. The association organizes an annual demonstration known as "Pasteur parlait français" at Pasteur Square on Rue Saint-Denis in Montreal. I attended the event in 2013, where Mr. Demers, then aged 98, delivered a speech in a remarkably firm voice, calling on French-speaking scientists to have their articles published in French-language journals. Pierre Demers died January 29, 2017, at the age of 102.

Contrary to what some have reported, Pierre Demers was not the only French Canadian to have worked at the Montreal Laboratory. Jacques Hébert (1923–2013) authored six Montreal Laboratory reports on plutonium and uranium-233 extraction. After the war, he was hired as a physics professor at the University of Ottawa, where he taught for over 40 years. There was also physicist Paul Lorrain, chemist Adrien Cambron, engineer Adrien Prévost, computers Gilberte Leroux and Fernande Rioux, and numerous administrative assistants.

Work Begins in Earnest

Despite the agreement signed by Roosevelt and Churchill in August 1943, there were still storm clouds hanging over the Montreal Laboratory. General Groves was dragging his feet, and the clauses set out in the agreement were still only empty promises. With the appointment of James Chadwick as Britain's representative in Washington, things started to look up for the Laboratory. Chadwick, who had discovered the existence of neutrons and been awarded the Nobel Prize for Physics in 1935, moved to Washington and became the liaison between the two countries. The first meeting between the esteemed physicist and General Groves took place not long after the Quebec Conference. On September 18, 1943, Groves headed up a U.S. delegation to inspect the Montreal Laboratory for the first time. The challenges were considerable.

Negotiations were required to lay the groundwork for collaboration between the two projects. Numerous meetings were held throughout the fall, but still no resources were freed up for the Montreal facility. The relationship between Chadwick and Groves proved to be decisive. Against all odds, Chadwick, an introverted and shy man, forged a friendship with General Groves. The two men couldn't have been more different: Groves would make on-the-spot decisions, tended to speak his mind, and was considered something of a bully, while Chadwick would gather and mull over all the relevant information, sometimes for days on end, before arriving at a decision, all while remaining unfailingly—sometimes, excessively—polite.

A strange sort of ballet played out at one crucial meeting of the Canada-United States-England tripartite committee on the future of the Montreal Laboratory. Before the meeting, a number of intense talks were held to clarify the goals of the research group and guarantee the Americans' contribution. The Canadian side included NRC director Mackenzie and, of course, Halban, although his position was somewhat precarious by this time. The Americans were represented by some heavy hitters, including a special advisor to Roosevelt and, inevitably, by General Groves. Chadwick was the British representative. The meeting got off to a disastrous start for the Montreal Laboratory when the Americans presented a draft agreement essentially maintaining the status quo, which would have sounded the death knell for the Canadian project. Chadwick immediately got up and went over to have a few quiet words with Groves. The two men left the room, leaving the others to wait nervously. After twenty minutes or so, Chadwick and Groves returned with a new agreement that all the parties quickly moved to accept. Chadwick had done more than just save the day; he had ensured the Montreal Laboratory's future, confirming its objective— the same one Halban had established from the outset—of developing a uranium/heavy water nuclear reactor as well as a method for separating plutonium. And so, the British team's initial plan to have one foot in the energy camp and the other in bomb development remained in place.

Finally, after so many twists and turns, in early 1944, the Quebec Agreement was honoured, and the Montreal Laboratory's work began in earnest. From that moment on, there were numerous back-and-forth trips between Montreal and Chicago. The researchers from Canada were given free access to the new Argonne campus some fifty kilometres west of the University of Chicago (the U.S. research team had outgrown the football stadium stands). The Montreal Laboratory's travel reports show that research groups from Montreal made nine separate return trips to Chicago between October 1943 and June 1944.[9] And that wasn't including the visits by American researchers to Montreal during the same period. Groves, however, refused the Montreal researchers access to the Manhattan Project's most top-secret sites, including Los Alamos, where the dif-ferent bomb models were tested.

Halban's fall from grace was a gradual process. It was decided he would be replaced as the director of the Montreal Laboratory, but it was a decision he only accepted grudgingly, and after some time. After all, he had been the man behind the research institute's founding. His initiative and willingness to accept risk had been outstripped by his inability to control his team. In December, Groves confirmed that he wanted Halban replaced by a Brit, and the name of one John Cockcroft came up. Cockcroft, who had been working on radar research, was unable to relocate right away to Montreal, and it wasn't until April 26, 1944, that he arrived in Quebec, where he spent his first day with Pierre Auger, the experimental physics group director, who laid out for him the team's technical and human resource problems of the past year and a half.

The following day, Halban introduced Cockcroft to the leaders of the different research groups, explaining that Cockcroft had come to lend a hand to manage the Laboratory's administration. Cockcroft calmly set the record straight, stating categorically that he was, in fact, the new director of the Laboratory, and announcing that he was putting Halban in charge of the experimental physics division,[10] as Auger was poised to leave Montreal, having been summoned back to France by General de Gaulle. Halban would occupy his new position in a mostly theoretical capacity, as the day-to-day management of the experimental physics group was actually overseen by Alan Nunn May, another British physicist we will discuss in greater detail in a later chapter. As Bertrand Goldschmidt put it, "Thus Halban had lost control of the management of the Montreal Laboratory, but with dignity."[11] From that moment on, the Laboratory grew significantly, launching one project after the other and, at one point, employing some 400 people.

Physicist Pierre Auger at the Montreal Laboratory in June 1944 ponders the future that awaits him in France as it is being liberated by the Allies. (Photographer unknown, National Research Council Canada, Cockcroft family photo album.)

Hans Halban was barred from returning to work as a physicist in France for eight years after the war ended. Halban is seen here at a congress in the late 1950s. (Photographer unknown, personal archives of Philippe Halban.)

JEANNE LECAINE-AGNEW
AND THE OTHER WOMEN
AT THE LABORATORY

The shortage of raw materials and the lack of clear direction didn't prevent some of the scientists from working. One such case in point, that of theoretician Jeanne LeCaine-Agnew, is particularly interesting as it gives insight into the work carried out in the shadows, and highlights the role women played in the Laboratory's work.

The contribution of these women has been largely ignored or even, dare I say, dismissed. Marianne Gosztonyi Ainley and Catherine Millar of the Simone-de-Beauvoir Institute at Concordia University, two feminists who have studied the role of women scientists in Canada in the 20th century, wrote in a 1991 article, that "In the 1930s, the University of Toronto trained and employed several competent women physicists, but the Atomic Energy Project, although it had thirty physicists, did not employ them."[12] Likewise, the official plaque at Université de Montréal commemorating atomic research and unveiled in 1962 by the Duke of Edinburgh, lists the names of male scientists only.

Jeanne LeCaine endured a similar lack of respect during her academic career. After earning a bachelor's degree, then a master's in mathematics in only four years at Queen's University in Kingston, Ontario, the director she hoped would supervise her PhD dissertation at Harvard was originally reluctant to take her on, claiming that his "previous female student had married and had five children."[13]

But LeCaine wasn't about to let the misogynistic views of the day hold her back, and she went on to complete her PhD in 1941 at Harvard, where she met her future husband, who was sent off to serve in the war. Jeanne LeCaine-Agnew, too, felt compelled to contribute to the war effort, and returned to Canada. She was hired by the National Research Council of Canada, in Ottawa, where her brother Hugh worked. After much insistence by George Laurence, her boss at NRC finally agreed to let her go,

Mathematician Jeanne LeCaine was reunited with her husband, Theodore Agnew, at the end of the war, after 27 months apart. (Photographer unknown, personal archives of the LeCaine-Agnew family.)

and after being interviewed by Placzek, she was transferred to the Tube Alloys project in Montreal. There, she shared a desk with Carson Mark, a young Canadian mathematician who was working on the theory of neutron transport under Placzek.

She was part of a small group of physicists and mathematicians who laid the groundwork for the physics of the nuclear reactors still in use today. Jeanne LeCaine co-authored, among other works, a report entitled "Elementary Approximation in the Theory of Neutron Diffusion."[14] The theoreticians were aided in their calculations—for instance, to determine the amount of heavy water and uranium required, and how each material would be arranged in the reactor—by a team of women known as "computers," a term mentioned earlier in the book. LeCaine Agnew felt an affinity for these women who would spend their days doing meticulous mathematical calculations. She grew especially fond of Joan Wilkie, and the two women remained lifelong friends.[15]

NEW ERA FOR THE LABORATORY

An Astute Shift to Energy

The MAUD Committee had been ahead of its time. As early as 1940, the group of scientists had rightly deduced that the future of nuclear science would split into two separate paths: military and civilian. But in 1942, the British realized they weren't the only ones seeking to build the bomb. Worse still, they had been shunted into a secondary role by the Americans. The outlook for nuclear as an energy source was therefore a more appealing option, especially considering that Hans Halban and his partner Lew Kowarski had continued their experiments all along to design a reactor using heavy water as a moderator. In fact, most of the employees at the Montreal Laboratory had been assigned to that precise task. Just because Halban had been sidelined, it didn't mean their mission had been, too. Quite the contrary...

John Cockcroft's arrival at the Laboratory in April 1944 signalled a wind of change. Cockcroft was well known—and, more importantly, well respected—among the scientists. Trained as an engineer, mathematician, and physicist, Cockcroft was another of Rutherford's protégés, having earned his PhD at the Cavendish Laboratory in 1925. His mathematical skills would serve him well in his career. In

The Cockcroft-Walton particle accelerator created at the Cavendish Laboratory in the 1930s earned John Cockcroft a reputation as a physicist before he was named director of the Montreal Laboratory in 1944. (Wikimedia Commons, user: Geni, 2012.)

the late 1920s, for example, he calculated the energy required for a proton to collide with and split the nucleus of an atom. His most well-known accomplishment was, however, the invention of the particle accelerator, which led to James Chadwick's discovery of the neutron.

The scientific community had already had the chance to appreciate the ingenuity of the two men now working together on the Montreal project. In the early days of the war, Cockcroft had been named assistant director of scientific research at England's Ministry of Supply, where he worked primarily on radar. He was also part of the MAUD Committee that paved the way for the country's atomic endeavours. In the wake of the Quebec Conference, in late 1943, Cockcroft was approached by Wallace Akers, director of the Tube Alloys project, to see if he would be willing to replace Hans Halban as the director of the Montreal Laboratory. Cockcroft accepted, and relocated his family (his wife Eunice Elizabeth and their five children) to Montreal. He arrived on his own in April 1944, settling in temporarily at the Ritz-Carlton Hotel. When his family joined him later that summer, they all moved into a large, timber-frame house in Westmount that offered breathtaking views over the city and the St. Lawrence River.

Cockcroft immediately set out to procure the additional staff and equipment the Laboratory needed. His plan was to design and build a large, heavy-water nuclear reactor, the NRX, an acronym for National Research eXperimental. The goal was to create cleaner energy than coal using the atom. At the time, smog episodes were becoming increasingly common in cities, and the thick haze overhead was endangering the health of their occupants. The city of St. Louis, Missouri, for instance, had to leave the city's

John Cockcroft took this photo of his five children (Thea, Jo, Elizabeth, Cathy, and Christopher) on the front steps of their home, shortly after they arrived in Montreal, in 1944. (John Cockcroft, Cockcroft family album.)

John Cockcroft and his daughter Joan Theodora (Thea) ready for a day of skiing in Sainte-Adèle, Quebec, 1944. (Photographer unknown, Cockcroft family album.)

streetlights on all day long during one such episode in 1939. That same year, the Joliot-Curie team put together a patent application on energy production using an atomic pile.

Meanwhile in Rome, Fermi and his "Via Panisperna Boys" had done the same. The hope was that nuclear energy would provide a solution to the coal pollution problem. Plus, it appeared that, with very small amounts of uranium, it was possible to produce large quantities of power. The energy generated by one kilo of uranium is equivalent to that produced by 20 tonnes of coal! But there were still many technical wrinkles to be ironed out before a nuclear reactor could be hooked up to the power grid.

Cockcroft and the Laboratory scientists faced a significant problem: No heavy-water reactor the size of a full-scale NRX had been built to date. It was akin to the Wright brothers attempting to assemble a bomber aircraft before building a glider! Much of the data required to build a large reactor was still either simply unknown or extremely imprecise. For example, no-one knew what the exact spacing should be between the uranium rods (which would determine the size of the reactor) and, consequently, the amount of heavy water and concrete that was needed. In early 1944, Alan Nunn May, one of the British physicists, proposed a solution. He suggested building a small, very low-power model reactor, so as to assess the dimensions and test different configurations.[1] It would be a sort of test run before the great NRX adventure. The proposal was passed on to Halban, who by then was director of the experimental physics group, and he recommended it to Cockcroft, who approved the plan on the spot, and set about recruiting a project leader for the mini reactor. Little did Halban suspect he was about to suffer another blow, and not the least! To fill the new

The Montreal Laboratory mathematicians, physicists, and chemists, June 1944. Although women made up more than a quarter of the Laboratory's employees—several physicists, chemists, and mathematicians among them—not a single one was invited to be in this official photo.

Standing: Allan Munn, Bertrand Goldschmidt, William Ozeroff, Bernice Sargent, Gordon Graham, Jules Guéron, Herbert Freundlich, Hans Halban, Ronald Newell, Frank Jackson, John Cockcroft, Pierre Auger, Stefan Bauer, Quentin Lawrence, Alan Nunn May. Sitting: John Knowles, Pierre Demers, James Leicester, Henry Seligman, Ernest Courant, Ted Hincks, Frederick Fenning, George Laurence, Bruno Pontecorvo, George Volkoff, Alvin Weinberg (U.S. liaison), Georg Placzek. (Personal archives of the Cockcroft family.)

position, the Laboratory required someone with a thorough understanding of neutron physics, advanced knowledge of engineering, as well as leadership skills. A few names were proposed by Halban, including his preferred choice, Stefan Bauer, an Austrian engineer who'd become a British citizen in 1942. Meanwhile, Cockcroft conducted his own search, consulted with the Laboratory's other division directors, and came back with his choice: Lew Kowarski, Halban's former research partner.

At a meeting in August 1944, Kowarski got the nod, and was transferred to Montreal, at Cockcroft's explicit request, to take on the position of project leader. The decision must have been a bitter pill for Halban to swallow, especially as he didn't consider Kowarski up to the task. In his diary, Halban methodically noted all the important events affecting the Laboratory, but of the arrival of his former assistant in late July 1944, not a word. His July 31 entry couldn't have been more succinct: "Cleaned up some old files." Several weeks later, Kowarski's wife, Dora, arrived in Montreal with their daughter, Irène, who was just shy of her 8th birthday. On August 25, 1944, the Kowarskis celebrated their only child's birthday in the garden of their new home on Mount Royal, quite close to St. Joseph's Oratory. Irène was taken aback when, in the middle of the party, several of the adults in attendance, including her mother, burst into tears. The Liberation of Paris had just been announced![2]

As soon as he was settled in at Université de Montréal, Lew Kowarski got down to work and formed a small team of eight people to design, build, and commission the Zero Energy Experimental Pile, or more readily pronounce-able ZEEP, as Kowarski suggested it be known. Bertrand Goldschmidt, who knew Kowarski from before the war at

Physicist Lew Kowarski, his wife Dora, and their daughter Irène rented a house on Mount Royal near St. Joseph's Oratory and Université de Montréal, 1944. (Photographer unknown, personal archives of Irène Kowarski.)

the Joliot-Curie Laboratory in Paris, was impressed by the physicist's newfound poise: "He no longer had anything in common with the badly groomed bear, surly and distant, whom I had known at the Curie laboratory eight years earlier. He had grown out of his old ways during his long stay in England […]. Fighting his unstable nature, he now seemed at ease, self-confident, and witty, full of paradoxes and puns. I had found it difficult before to believe the influence he had exercised on the Cambridge team in 1942, but now I understood why his colleagues had threatened to strike when they were transferred to Canada without him during his break with Halban."[3] When he arrived in Montreal, Kowarski was briefed by a physicist who introduced him to the fundamental ideas of pile structure that he had gathered on his visits to Chicago. According to Kowarski, he delivered the information in a sort of lecture in a single afternoon, which was quite enough![4]

The ZEEP design was relatively straightforward. Since it was a very low-power reactor, there was no need for a cooling system or elaborate shielding. In collaboration with the engineers, Kowarski opted for a classic shape, with a vertical metal cylinder in which uranium rods were inserted. Heavy water was then added, pumped in as you would if you were filling a swimming pool. In some ways, ZEEP was like a giant version of the barrels of heavy water that Halban and Kowarski had transported from Paris to Bordeaux, then to Cambridge, and, finally, to Montreal. Despite complaining they needed it more than the Montreal group did, the Americans reluctantly agreed to lend the team the heavy water they needed to achieve criticality. During the summer of 1945, Kowarski reached out to General Groves for the first time. Without further ado, Groves said to Kowarski, "Are you the man who is building this damn fool unnecessary experimental reactor?" "I am." "Well, America gives most of the heavy water for it, and it's very, very costly stuff. You see that you don't squander it."[5]

One key question was immediately raised: Where would ZEEP and its big brother NRX—the two reactors under development at the Montreal Laboratory—be built? Although a number of sites in Quebec and Ontario were proposed, the government chose instead another site altogether, near a military base some 200 kilometres north of Ottawa, in Chalk River, on the Ottawa River. And so, a brand-new laboratory (Chalk River) was built, along with a city to accommodate all the workers (Deep River), and it is still one of the largest research sites in Canada. In the beginning, the site was home only to a forest and several cottages along the riverbanks, which were promptly expropriated by the government. Defence Industries Limited (DIL) was selected as the main building contractor. You will recall that DIL owned Canada's biggest munitions factory, in Verdun, on the Island of Montreal. DIL, in turn, hired Fraser Brace, an American company whose Canadian subsidiary was headquartered in Montreal, to build the laboratory facility and employee lodgings. While a handful of scientists, engineers, and workers took up residence there in the summer of 1944, most of the Montreal Laboratory employees didn't relocate there until the fall of 1945, while John Cockcroft moved to Deep River with his family in November 1945.

The team Kowarski put together and supervised achieved a remarkable feat by completing the detailed conception, construction, and commissioning of a nuclear reactor in the space of only twelve months. To achieve criticality, heavy water was pumped into the tank around the uranium rods.

The first nuclear reactor outside the United States, ZEEP, seen here on June 21, 1945 (white building in the foreground). In the background, the larger NRX reactor, also under construction, was commissioned in 1947. (National Research Council of Canada.)

As the tank filled, more and more neutrons were slowed down (this was the role of the heavy water) and could subsequently be absorbed by uranium-235 atoms, causing them to fission and release heat. In more powerful reactors, this heat is used to boil water (similar to coal or gas power plants), and the steam that is generated activates a turbine that produces electricity. George Volkoff, the leader of the theoretical physics group, was in charge of calculating the critical height of the heavy water—the point at which a self-sustained chain reaction would occur—which he accomplished with the assistance of another Canadian physicist, John Stewart.

Within twelve months, all the preparation was complete, and the Chalk River ZEEP went critical on September 5, 1945, making Canada only the second country in the world, after the United States, to have successfully built a nuclear reactor. ZEEP provided many years of service to Canada's nuclear industry, until it was eventually decommissioned in 1968.

ZEEP operated virtually non-stop until 1947, with the experiments conducted on it aiding in the conception of the NRX. When the NRX reactor was commissioned, the heavy water from ZEEP was transferred to the larger reactor, signaling an end to the first phase of ZEEP operation.

Montreal and the Bomb

The Montreal Laboratory hadn't given up on its plan to participate in the race to build a bomb, even if England had come to the realization that it wouldn't be able to go it alone. The government was looking ahead to the post-war period now, and felt it unwise to leave the atomic monopoly to the United States.

Two elements with potential military applications— polonium and plutonium—were handled in Montreal. Research into polonium had begun during the troubled period of 1943 and early 1944, before Cockcroft arrived. In the fall of 1943, the Americans invited Goldschmidt and Paneth (director of the chemistry division) to Chicago to discuss this novel chemical element. Polonium is a radio-active metal (Po on the periodic table of elements) that was discovered by Marie and Pierre Curie at the turn of the century. It is found in microscopic quantities in nature, in rocks containing uranium and radium. In the 1930s, polonium took on a surprising new importance. Why? Because

HALBAN AND THE JOLIOT-CURIE COUPLE

Many references are made to Hans Halban's personality in his colleagues' anecdotes about him. Yet, Halban was also a man who went out of his way to help his family and friends who had stayed behind in France during the war. He remained especially close to Irène and Frédéric Joliot-Curie, providing them with financial and material assistance. A previously unreleased postcard Irène Joliot-Curie sent to Halban in Montreal bears witness to their friendship.

The card reads [our translation]: *"My dear Halban, we have received several more packages from you, always much appreciated. We will be taking a 15-day trip to the USSR. I expect it will be very interesting, although quite tiring. On the whole, we have had quite a pleasant season, and on Sundays, our garden is often teeming with friends. Speaking of which, if you could send us a few tennis balls, we would be most delighted. We have started playing again, with balls that date back to Antiquity. Warmest regards, Irène Joliot-Curie."*

Irène Joliot-Curie sent this postcard from Paris on June 10, 1945, to her friend Hans Halban in Montreal, in which she asks him to send her some new tennis balls! (Photo by Philippe Halban.)

Mon cher Halban,

nous avons encore reçu quelques petits colis de vous, toujours bien utiles.

Nous allons faire un voyage de 15 jours en URSS. Je pense que ce sera très intéressant, mais assez fatigant.

Nous avons eu dans l'ensemble une saison bien agréable cette année et notre maison, notre jardin sont envahis le Dimanche par de nombreux amis. ...

Six months earlier, in December 1944, Halban had requested, and obtained, permission from his British superiors to travel to France to visit Frédéric Joliot. Halban was keen to see his former boss of the Curie Laboratory again, to bring him up to speed on the progress of their atomic research during the war. Halban—and, in fact, all those who worked at the Montreal Laboratory—was forbidden from speaking about his work with anyone outside the project so, before he could go, he had to clarify with the British which topics he was allowed to discuss with Joliot. Judging by Halban's diary, his December 1944 meeting with Joliot in Paris was something of a letdown. Halban could hardly get a word in edgewise, as Joliot talked nonstop about his role in the French Resistance. Halban returned to Montreal somewhat disappointed, but happy to have accomplished his duty to his adopted country.

Unfortunately for Halban, the Americans hadn't been made aware of his visit, and were furious when they learned of the meeting. It confirmed their worse fears that the expatriates at the Montreal Laboratory would start sharing top-secret information at the first opportunity. This incident would come back to haunt Halban later when he attempted to find a position worthy of his qualifications, and when the war ended, he wasn't called back to Collège de France as he had been hoping. Instead, he found a position at Oxford University as the head of a research group connected to the Harwell Laboratory (the British post-war equivalent of the Montreal Laboratory). It took another eight years before Halban was invited, in 1954, by the leader of the French government, Pierre Mendès-France, to head up a new nuclear research laboratory in Saclay, just outside Paris. The research centre was responsible for coordinating France's nuclear energy research as well as its military program to develop an atomic bomb. Halban later developed health problems that forced him to retire in 1961. He died three years later, in 1964, following an operation at the American Hospital of Paris, leaving behind his third wife, Micheline Lazard-Vernier, and his three children (Mauld, Pierre, and Philippe).

when a source of polonium (which emits alpha-particles) is combined with beryllium, the beryllium releases neutrons. This combination of polonium and beryllium was what all the atomic physics researchers were using in the late 1930s. The Montreal Laboratory had in its possession a small but carefully preserved amount of polonium-beryllium. Without revealing all the details of the process to Goldschmidt and Paneth, the Americans did, however, let on that they were looking for a large quantity of polonium. The two scientists were very familiar with the element, and the Americans wanted to pick their brains. It didn't take Goldschmidt and Paneth long to deduce that the plan was to use the polonium as an initiator for a plutonium bomb.

And so, in October 1943, Goldschmidt flew once again to New York City, where he worked at Memorial Hospital to recover the polonium contained in the sources of radium used for treating cancer. In his words: "The delicate manipulation led to the isolation of the largest quantity of polonium ever prepared up to then. As I had been ordered, I brought it back immediately from New York to Montreal where, less than twenty-four hours later, a U.S. officer—who had also come from New York during a snow storm—took possession of it to transport it to Los Alamos."[6] This was something of a political move: The Canadians and British wanted it known that the polonium was the result of an official collaboration between the Montreal Laboratory and the Manhattan Project, hence the stopover in Montreal before the material was shipped to Los Alamos, where it would be used in research on how to detonate a plutonium-239 atomic bomb. But the supply was soon exhausted, and the Americans had to find a new source elsewhere.

In the agreement Chadwick had worked out with Groves, the Montreal Laboratory was allowed to work on another military aspect: plutonium separation. As we have seen, there are two types of atomic bomb—one using uranium-235, and the other, plutonium-239. Building a plutonium bomb meant first building a nuclear reactor in which uranium rods would be irradiated, then separating the plutonium that was produced as a by-product inside the now-highly radioactive rods. While the team knew that a reactor powerful enough to produce enough plutonium (the NRX reactor) would not be built during the war, they nonetheless worked on a plutonium separation method so they'd be ready once the NRX was commissioned.

One major thorn in the side of the Montreal Laboratory was the fact that General Groves had forbidden any exchange of information about the discoveries made by the Manhattan Project. Yet, the Americans were much further ahead in this regard, and in 1944, they already had reactors operating in Hanford, Washington, to produce plutonium. Montreal was therefore forced to virtually start from scratch, since the U.S. bomb project was not dependent on the plutonium that would be produced in Canada. The only concession Chadwick was able to obtain from Groves was the promise of two irradiated rods from a nuclear reactor in operation in Oak Ridge, Tennessee. Those rods contained just a few grams of plutonium—a priceless treasure for Goldschmidt's group. One day in July 1944, Goldschmidt and Jules Guéron (another French chemist in charge of a research group in Montreal) got a call from the RCMP officer posted at the Laboratory entrance: a package had just arrived for them from the United States. In fact, it was two cars driven by armed guards, and each of the vehicles was carrying an irradiated uranium rod housed in a lead "flask." The rods had been secretly transported across Tennessee, Virginia, Maryland, Pennsylvania, and New York, before crossing into Quebec—a journey of over 1,600 km.

The Montreal chemists immediately set about testing different methods to separate the plutonium from its uranium gangue. The presence of highly radioactive fissile material in the uranium rods complicated the operation enormously. Goldschmidt's team's goal was to develop a plutonium extraction method that would work in an average-size plant, not just in the lab. This meant the process had to be relatively simple and easy to replicate. After removing the aluminum sheath from around the irradiated uranium rods, the rods were dissolved in nitric acid, creating plutonium and uranium nitrates in a liquid solution containing all the fission products. The next step was more tedious, as it required testing some 200 solvents to determine which was the most effective at extracting the plutonium. The solvents had to be immiscible in water, in other words, like olive oil, they wouldn't mix with water. Each solvent had to be tested in a flask, at different temperatures, to see which it would dissolve most effectively: the uranium, the plutonium, or the fission products. It was a task that required many long hours in the lab. The mixtures had to be shaken by hand. Then, the solvent was extracted, and its level of radioactivity measured.

The work was not without risk. In his autobiography, John Spinks, another chemist who was part of Goldschmidt's group, recounts, "At one stage when we had dissolved up the slug and had successfully removed the greater part of uranium, the plutonium seemed to disappear. It was found adsorbed (or stuck) on a few flakes of white material which had dissolved out of the stainless steel from which the container was made. That was bad enough, but practically all the fission product activity had been adsorbed on the same material, and separating them was a nightmare."[7] In summer 1945, the researchers finally identified the formulation that worked best: it was a product used in the manufacture of rubber—triglycol dichloride, dubbed "trigly." The plutonium extraction process was later improved by a group of NRC scientists in Ottawa, and it was this process that would be used in Chalk River when the NRX was commissioned in 1947.

All of these research programs required the handling of highly radioactive—and therefore highly dangerous—products. To see its projects through to fruition and protect its staff at the same time, the Montreal Laboratory would have to adapt.

THE DE GAULLE MEETING

In June 1944, some of the French scientists working at the Laboratory (Auger, Goldschmidt, and Guéron) learned that the leader of Free France, General de Gaulle, was scheduled to make a visit to Ottawa in the near future. Knowing how hostile the Americans were to any foreign participation in the Manhattan Project, the French expatriates knew full well that, once the war was over, the United States would seek to retain full control, for as long as possible, over the monopoly of knowledge about atomic energy and the bomb. As a result, they decided to break their oath of secrecy and inform de Gaulle of the objective of the atomic research being conducted.

They managed to convince Gabriel Bonneau, Free France's representative in Canada, to allow them a ten-minute private meeting with the general for a secret communication of the utmost importance. Bonneau had just one condition: only one person could speak to de Gaulle. Guéron was the trio's choice, because he knew the general and, moreover, since he was being paid by Free France, he hadn't signed the *Official Secrets Act*. On July 11, 1944, after giving a speech in front of the Parliament on France's post-war participation in international peace and cooperation, the general went to the building where the French delegation was being housed. Having been informed of the meeting in advance, he announced he was going to wash his hands, and headed to the bathroom at the end of the hall. But instead of going into the bathroom, he opened a door to another room on the opposite side of the hall, where Guéron was waiting for him. The scientist quickly delivered the message he, Auger, and Goldschmidt had agreed on: "A weapon of extraordinary power, based on uranium, should be ready in one year and be used first against Japan. The possession of the weapon, perfected in the United States, was going to give that country a considerable advantage in the world after the war. It was absolutely necessary to resume atomic research in France as rapidly as possible. Joliot and Francis Perrin were the ones who would have to organize this task."[8]

De Gaulle addressing the crowd on Parliament Hill in Ottawa on July 11, 1944, one month before the liberation of Paris and his return to France. (Photographer unknown; Library and Archives Canada C-026947.)

The general kept Guéron's message in mind, but for the moment, he had more pressing concerns. The Normandy landing had taken place on June 6, and the Allied troops were in the process of taking back France from the Germans. Paris was liberated on August 25. Five days earlier, de Gaulle had tapped Frédéric Joliot to head up France's National Centre for Scientific Research (*Centre National de la Recherche Scientifique*).

In 1945, Joliot helped found the country's Atomic Energy Commission (*Commissariat à l'énergie atomique*), and was appointed its first High Commissioner for Atomic Energy.

THE STORY OF ALMA CHACKETT

Among the chemists who worked in Montreal was a British woman, Alma Chackett, who was still alive and active, at age 102, at the time this book was written. I had the opportunity to meet with her, and we had several long conversations. Here is her story, as she described it to me:

Born Alma Thompson, she studied chemistry at the University of Birmingham, where she earned a bachelor's degree in 1940. There, Thompson met her future husband, Ken Chackett, also a chemistry student. In the 1940s, the University of Birmingham was home to one of the most advanced atomic research groups in the world. It was where, in 1940, Otto Frisch and Rudolf Peierls wrote their famous memorandum detailing the critical mass of uranium-235 required to build a nuclear bomb. After graduating, Alma worked first for the Northern Aluminium Company. Together with her colleagues, she was tasked with testing the composition of alloys used to make parts for Lancaster bombers and Spitfire and Hurricane fighter planes. She worked long hours, because the British industry had rallied all its resources to build aircraft in preparation for the Battle of Britain. Achieving the proper composition of the alloys that would be used in these parts was crucial to the durability of the country's warplanes.

In the summer of 1944, Ken Chackett completed his PhD at the University of Birmingham on techniques for separating noble gases (helium, neon, argon, etc.). He was recruited by Professor Fritz Paneth to work on the Tube Alloys project in Montreal. Ken and Alma were married on July 10, 1944 and, only three weeks later, Ken boarded the *Queen Mary*, bound for New York City, where he took the train to Montreal.

Officially, Alma was not allowed to be employed by the government on the same project as her husband. She hoped, however, that she would be given permission to join him later. She eventually received a letter summoning her to the Canadian Embassy in London's Trafalgar Square. She was questioned at length, and had to undergo a medical exam. During the interview, a V-1 missile fell nearby, and while the others in the room dove for cover beneath their desks, Alma remained calmly seated on her chair.

Not long after she returned to Birmingham, Alma received another letter ordering her to pack her bags and head to a specific platform at the New Street railway station, where she was to take train W17. At the station, a W17 sticker was affixed to her bags, and she boarded the train. She had no idea where she was headed, but concluded from the passing landscape that she was on her way to Liverpool. After stopping there, the train continued on to its final destination: Glasgow, Scotland. From there, Alma boarded the *Aquitania* en route for Halifax via St. John's, Newfoundland. Upon her arrival in Montreal, she was immediately hired by the National Research Council of Canada.

Alma Chackett's name appears on a list of Montreal Laboratory staff, dated May 1945, as a "locally engaged" staff member with an annual salary of $1,680.[9] Her husband was hired by the Laboratory in August 1944 as a junior scientist, and earned an annual salary of £330, which worked out to roughly $1,460 at the time.

In other words, Alma was paid more than her husband, which was highly unusual for those days.[10]

The chemists working on the Tube Alloys project were divided into small groups, with each group studying the properties of different uranium fission products. Alma studied bromine and iodine under Jules Guéron, while Ken's focus was on the noble gases neon, argon, and xenon. Alma noted that the scientists tended not to mingle socially with the other employees. Of course, due to the secret nature of the project, they were not encouraged to discuss their work. What's more, the scientific research they were conducting did not lend itself well to "regular" hours. The employees would take their breaks when they could, and would often go out to grab a quick bite to eat nearby.

Chemist Alma Chackett enjoying her newly built, spacious apartment close to Université de Montréal, 1945. (Ken Chackett, personal archives of Alma Chackett.)

While Ken and Alma worked very long hours, their life in Canada was pleasant compared to the dangers and hardships they had endured in England. Alma recalled the couple getting away for a holiday in the Laurentian Mountains north of Montreal, where they stayed in a log cabin with a group of friends and admired the colourful fall foliage as they paddled canoes on a nearby lake. From time to time, the Laboratory's scientists were invited to stay at a large house in Beaurepaire (a village now merged with the City of Beaconsfield on the Island of Montreal).

The couple also visited some cousins in Toronto and went to Niagara Falls over the Christmas holidays in 1945. It was a happy time for the Chacketts, although there was always the nagging worry for the well-being of their loved ones in England. And in fact, much of their free time was devoted to putting together and sending care packages to their families back home.

In the summer of 1945, the Montreal Laboratory's chemistry group posed for a photograph. Alma Chackett kept a copy of

From time to time, the Montreal Laboratory employees had the chance to get out of the city and stay at a guest house in Beaurepaire (today, part of Beaconsfield), on the Island of Montreal. (Ken Chackett, personal archives of Alma Chackett.)

the photo (see below). It is a remarkable memento, as it is one of only three known photos of the Montreal Laboratory group, and one of only two that include the female members of the group. We were able to identify 25 of the people. One-third of the 42 individuals in the photo are women.

Ken and Alma's Montreal adventure ended in Summer 1946 when Ken rejoined Professor Paneth at Durham University in the U.K. and Alma was hired by the university observatory. The couple moved into the house adjacent to the observatory, so she could conduct weather recordings there. It was there that the couple's two daughters, Lesley and Daphne, were born. In 1952, the family returned to the University of Birmingham (where the couple had met). Ken was hired as a researcher, and he supervised doctoral students. Alma joined him there, and their team conducted ground-breaking research determining the half-life of radioactive isotopes. Their lab housed the University of Birmingham Physics Department's first cyclotron. It was in operation from 1948 to 1999. Alma Chackett worked as a researcher in the physics department, where she used the cyclotron right up until 1980. During that time, she co-authored, together with her husband and several other researchers, a dozen scientific articles. Ken and Alma Chackett retired together in 1980. Ken passed away in 2013 after a long illness, one year before the couple would have celebrated their 70th wedding anniversary. At the venerable age of 102, Alma is still active. She lives in her own home and tends her garden.

In this photo of the chemistry division of the Montreal Laboratory taken in Summer 1945, some of the scientists who appear in it include Ken and Alma Chackett, Bertrand Goldschmidt, Fritz Paneth, and Frank Morgan. (National Research Council of Canada, personal archives of Alma Chackett.) 1: Kenneth Musgrave; 2: Bertrand Goldschmidt; 3: Albert English; 4: Alma Chackett; 5: Geoffrey Wilkinson (future Nobel Prize Winner in Chemistry); 6: Kenneth Chackett; 7: Henry Heal; 8: William Grummitt; 9: Jack Sutton; 11: Frank Morgan; 12: Alan Vroom; 13: Maurice Lister; 14: Fritz Paneth (director of the chemistry division); 15: Leslie Cook; 16: Ruth Golfman; 17: Graham Martin; 18: Leo Yaffe; 19: Robert Betts; 20: Allan Lloyd Thompson; 21: Ethel Kerr; 22: Jules Guéron; 23: Samuel Epstein; 24: Margaret Kingdon; 25: Patricia Gorie; 26: Gerda Leicester.

Portrait of Edgar Gariépy, whose photos of the Montreal Laboratory mysteriously disappeared. (City of Montreal Archives.)

THE MYSTERIOUS DISAPPEARANCE
OF EDGAR GARIÉPY'S PHOTOS

There are very few remaining photographs of the Montreal Laboratory, and most of those are stored away in the family photo albums of its former employees. However, there was a well-known professional Quebec photographer who was commissioned by the Montreal Laboratory. Edgar Gariépy, whose photos are considered today a remarkable source of images of Quebec from the first half of the 20th century,[11] appears on the Laboratory's list of employees from Summer 1944 until the end of 1945. Surely, during those eighteen months, he must have taken dozens, if not hundreds, of photos. Otherwise, why would he have been hired? Those invaluable historic photos appear to have been lost. Only the two group portraits on pages 69 and 104 were likely taken by Gariépy, as Alma Chackett mentioned in our interviews.

The Edgar Gariépy archives, which contain thousands of photographs, are held by the City of Montreal. Despite extensive searching of this fonds, as well as at Archives nationales du Québec, I found nothing in connection to the Montreal Laboratory. These mysterious photographs are nowhere to be found in the UK National Archives either, or even at the National Research Council of Canada archives I visited in Ottawa. Perhaps they are buried somewhere within Library and Archives Canada's labyrinthine system, but there again, I was unable to locate a single one. If ever those "lost" photos do reappear, they will serve to safeguard the memory of the Montreal Laboratory beyond the written word.

RADIOACTIVE CONTAMINATION AND ACCIDENTS

John Spinks was a young British chemist who worked under Bertrand Goldschmidt to separate the plutonium contained in the irradiated rods produced at Oak Ridge. After completing his PhD at Cambridge, Spinks was hired as a professor of chemistry by the University of Saskatchewan, in Saskatoon. That's where he met his future wife, Mary Strelioff, a farmer's daughter, whom he married in 1939. One day in 1944, when he got home from his workday at the Montreal Laboratory, Mary said to him: "John, your neck is red—I wish you would get rid of this silly habit of scratching your neck." He replied that he hadn't been scratching his neck; that it had probably just been a scratchy collar. "Then change your shirt!" she replied. In a few days, the

Chemist John Spinks discovered that one of the chemistry labs had been contaminated with radioactive products. (Local History Room, Saskatoon Public Library.)

redness disappeared. Two or three weeks later, the same thing happened again and his wife remarked that it was odd that it was the same shirt.

That evening, John took the shirt to the lab and tested it with a Geiger counter. In his words, "it was lousy with radioactivity—evidently a spot of highly active material had been splashed on my collar. Quite obviously, splashes might have gone in other places and would have dried off without leaving even a stain. On testing the floor of the lab, I found it was highly radioactive, and so were the walls and the ceiling, as a result of radioactive dust."[12] The next morning, Spinks reported his findings to his boss, Bertrand Goldschmidt. One of the team's physicists, who happened to be in the room at the time, did some tests, too, and confirmed Spinks' report. He immediately telephoned Cockcroft, telling him to bar the chemists from entering any other part of the building, for fear they would contaminate the entire place.

Spinks was immediately appointed to sit on the newly created Radiation Safety Committee. He was tasked with conducting a general contamination survey of the laboratory. This study is documented in the CI-73 report entitled "Contamination in the active laboratories."[13] The report is only seven pages, including three pages of charts, but it makes for fascinating reading. It notes that one of the rooms used to separate the plutonium contained several highly radioactive spots, particularly the vapour extraction hood and several of the work tables. It appeared the contamination stemmed mostly from small amounts of liquid radioactive product that had spilled on the floor and not been cleaned up properly. The droplets stuck to the soles of the scientists' shoes, and were likely transported that way to other parts of the lab. Spinks discovered that Bertrand Goldschmidt's shoes were especially contaminated, and they were confiscated as radioactive waste. Goldschmidt demanded compensation, but since there were no provisions for this kind of expense claim, his request had to be passed up the chain, starting with Professor Paneth, then Dr. Cockcroft, then Dr. Edgar Steacie (director of the National Research Council of Canada Chemistry Division), who passed it on to Chalmers Jack Mackenzie (president of NRC), who forwarded it to Minister C. D. Howe, who finally approved the request and sent it back down the line. According to Goldschmidt, the request even made its way all the way up to Prime Minister Mackenzie King, although there is no document proving this![14]

There is also the oft-told story—albeit long after the fact—of Alfred "Alfie" Maddock, who first worked on the Tube Alloys project in Cambridge, where he was in charge of the supply of heavy water brought from Paris by Halban and Kowarski. He loved to play up his role as the mad scientist who accidentally blew up all the equipment he'd been using to conduct his PhD experiments. In Montreal, Maddock was part of Bertrand Goldschmidt's group working on plutonium separation. Legend has it that one night, as he was working on the highly radio-active material that had been extracted from the Oak Ridge irradiated rods, Maddock accidentally spilled the solution all over his wooden worktable. Desperate to recover the pluto-nium, he sawed a chunk out of the table and burned it, so he could collect the ashes containing the plutonium. As the story goes, he managed to recover 95 percent of the product. What happened to the remaining 5 percent is anyone's guess, but it can't have helped the Lab's decontamination efforts!

These incidents led to several recommendations being made, including regular readings of radioactivity levels, to ensure clean-up operations were effective and that there was no new contamination. At the Montreal Laboratory, employee health was taken seriously, as the dangers of radioactivity were well known by that time. Towards the end of 1944, a medical division was also added at the Laboratory.

The medical division's monthly reports made mention of cases in which worker health was impacted by radiation. Most of the employees had their blood tested to check their white blood cell counts. One of the first signs of radiation sickness is a drop in the number of white blood cells and, at the time, blood testing was the only known method for detecting con-tamination. Cockcroft, the Laboratory's director, was concerned about possible contamination, and he decided to considerably expand the Lab's research on employee health and the impact of radiation.

TWO CHEMISTS DOOMED TO AN EARLY DEMISE...

An intriguing, though as yet unconfirmed, story was brought to light by physicist Henry Duckworth. During the war, he worked at NRC in Ottawa and had the opportunity to visit the Montreal Laboratory. He recounts what he saw there in his autobiog-raphy, *One Version of the Facts: My Life in the Ivory Tower*:

> I visited the Montreal Laboratory for a two-day period during desperately cold weather in January 1945. [...] The radioactivity section of the Montreal Laboratory was directed by an old-time

radio-chemist, Friedrich Paneth [...]. In his pre-war career, he had never dealt with any but minute quantities of radioactive material. On that account, he was unaccustomed to taking the precautions required for more active samples. The consequence of this lack of foresight had struck the Laboratory immediately prior to my visit. It had just been realized that two of Paneth's younger staff, Heal and Morgan, had been exposed to fatal doses of radiation. They would not die instantly, but it was inevitable. Colleagues rushed to open doors for the doomed pair and, in other ways, tried to make their lives more bearable. I never heard how long they lived, but, in 1955, [...] I was startled to discover that Morgan was chairing a colloquium. I asked how Heal was and was told that he was well at last hearing. Thus, the two defied predictions [...] and received their death tributes long before their actual demises.[15]

It is, admittedly, a compelling story, and it provides a good deal of detail that makes it entirely plausible. I must say I was stunned when I first heard it. Henry Heal and Frank Morgan were, in fact, both chemists who worked with Fritz Paneth's group at the Montreal Laboratory. In 1945, they were age 25 and 24, respectively. Both men survived more than forty years after the incident reported by Duckworth, if it did, in fact, happen. Duckworth's description of Fritz Paneth is rather unflattering, and is not corroborated at all by those who worked with the man directly. Alma Chackett's daughter, Daphne MacDonagh, recounts a discussion she had with her mother about the story:

She (Alma Chackett) was quite angered by Henry Duckworth's story, and didn't believe a word of it. She was convinced that if the incident had actually occurred, she would have discussed it with my father, and would have remembered it. In any case, she didn't recall Professor Paneth being directly responsible for safety measures at the lab. There were radiation protection officers whose job it was to ensure the employees' safety, and she pointed out that everything was strictly monitored. If anyone was exposed to a dose even slightly above normal, they were removed from the job for a certain period. She feared the story was simply an attempt to discredit Professor Paneth. Why? One can only speculate. Frank Morgan, one of the two supposed victims, was known as a laid-back and happy man. He never behaved as if he had been handed a death sentence and, in fact, he went on to have four children.

Fortunately, the two men lived for many more years, suggesting that the incident either never happened, or was somewhat blown out of proportion.

I had the opportunity to reach out to the families of the two chemists, both of whom are deceased. None of their relatives had heard the story told by Henry Duckworth. That said, having worked in the chemistry division in charge of plutonium extraction, the two men would have most certainly been exposed to levels of radiation higher than today's standards. Towards the end of his life, Frank Morgan penned his autobiography, which hasn't been published to date. In it, he makes no mention of any event that could have exposed him to a lethal dose of radiation. At best, he refers to a journey in a van from Montreal to Chalk River to transport 5 mg of plutonium, accompanied by armed RCMP officers.[16] In my view, all the material handling that went on at the Laboratory was far more dangerous to the scientists' health than any radiation they may have been exposed to during that trip.

In early 2016, I contacted Janet Morgan, the daughter of Frank Morgan and Sheila Sadler, who was born in Montreal in December 1945. After reading the story of the foretold death of her father in 1945, she wrote: "Thanks for one of the most fascinating emails I've ever received. I am Frank Morgan's daughter (born in Dec 1945 in spite of everything) .... My parents had been warned in January 1945 not to conceive a child for at least two years, but that they clearly disregarded this advice." This strange story was never confirmed. And while Duckworth's story strikes me as an exaggeration—otherwise, how would the two men have survived more than forty more years?—I get the sense that there must have been some kind of incident during which they received a higher-than-normal dose of radiation. This would explain Duckworth's recollection and their relatively premature deaths in their sixties, one from cancer and the other from a heart attack.

The Cobalt "Bomb"

These contaminations would have very tangible repercussions. Cockcroft decided to hire a British physician and cancer specialist, Joseph Stanley Mitchell, to oversee the health of his staff. Mitchell was appointed to the recently created medicine and biology division, which focused on protecting the Laboratory employees against radiation, as well as conducting research into the fight against cancer, a field in which a fundamental discovery was soon to be made in Montreal.

In the early days of the war, Mitchell worked as a radiotherapist at the Emergency Medical Service in Cambridge. In 1944, when he joined the Montreal Laboratory, he brought with him a unique skillset, thanks to his medical degree and hands-on experience in radiology. Upon his arrival in Montreal, he immediately began searching for radioactive products that could serve as an alternative to radium in cancer treatments. Radium, which had been used since the early 20th century, had several major drawbacks. Joseph Mitchell was well aware of this, having used it to treat his patients. Up until then, treatment had consisted of injecting radium through the skin and directly into the tumour, however, this only worked if the tumour was close to the surface and not lodged in an internal organ. To treat skin cancers, radium poultices were applied directly to the tumour.

Mitchell and his collaborators studied the properties of fission products and natural elements that could be irradiated in a nuclear reactor to replace the radium with a more effective atom. Their research paid off. At Cockcroft's request, in 1945, Mitchell summarized his research in an HI-15 report entitled "Applications of recent advances in nuclear physics to medicine."[17] The report was officially published in the *British Journal of Cancer* in 1947.

Mitchell recommended using cobalt-60 to fight cancer for a number of reasons. First, because the gamma rays it emits penetrate to a much greater degree than the alpha particles emitted by radium. This meant cancers located in internal organs could be treated. The half-life of cobalt-60 was also ideal: not too long, yet not too short. Thanks to this property, hospitals could store it, and it would remain active for about fifteen years. Another major advantage was that it was much cheaper to produce than radium. In the 1930s, one gram of radium cost anywhere between CAN$500,000 to CAN$1 million. Mitchell calculated that the nuclear reactor under construction—the NRX—could produce several hundred curies of radiocobalt per month, ensuring a significant supply on a regular basis.

The story of cobalt-60, which took root in Montreal, continued on in Chalk River, thanks to a new recruit. In May 1945, André Cipriani joined the medicine and biology division of the Montreal Laboratory. Born in Port-of-Spain, Trinidad, Cipriani studied physics, engineering, and medicine at McGill University. While in Montreal, Cipriani, too, conducted basic research on cobalt-60, measuring the energy from gamma rays emitted by the cobalt—a crucial measurement to enable it to be used in the medical field. Cobalt was placed in a shielded device next to a seated or supine patient, and the gamma rays it emitted were directed at the malignant tumour.

In the fall of 1945, André Cipriani moved to Chalk River, where he was appointed director of the biology division. The following summer, one of his colleagues gave a two-week course on radiotherapy. One of the attendees was Harold

Johns, a physicist from the University of Saskatchewan. In 1949, scientists tried putting natural cobalt rods, i.e., cobalt-59 into the NRX so they would absorb neutrons and transform into cobalt-60. The cobalt produced this way cost 6,000 times less than radium, and its rays penetrated deeper. The first rods of irradiated cobalt-60 were sold to the University of Saskatchewan and to Eldorado Mining and Refining, a crown corporation that controlled the extraction and processing of uranium in Canada.

At the University of Saskatchewan, Harold Johns, assisted by his doctoral students, designed a "cobalt bomb," which is how they referred to cancer treatment machines in those days. The premier of Saskatchewan, Tommy Douglas, renowned for introducing the first universal healthcare insurance in a Canadian province, in 1943, gave the green light to the university to purchase cobalt-60 from Chalk River. The cobalt was delivered in late July 1951. Extreme care had to be taken to avoid people being accidentally irradiated or exposed for excessively long periods. Direct exposure of only a few minutes was enough to be lethal. The first patient in Saskatchewan—in fact, in the world—to receive cobalt therapy was a 43-year-old mother of four who was treated for cervical cancer in November 1951. The treatment worked, and she went on to live for another 47 years.

At the same time, the Eldorado Company built a similar machine in London, Ontario, which became operational a few days after the one in Saskatoon. With that, Canada became the world's leading producer of cancer treatment machines. From 1950 to 2000, the country supplied over 50 percent of the "cobalt bombs" used in 80 different countries. It is estimated that, during that period, seven million patients were treated using cobalt-60 in Canadian machines.

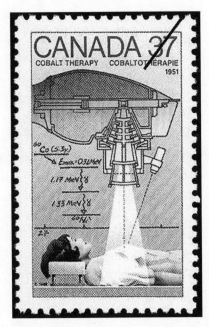

Stamp commemorating the first cobalt-60 treatments, in 1951.
(Canada Post.)

So, while it may not have built an atomic bomb, the Montreal
Laboratory did design a "cobalt bomb" that could be used to
treat cancer patients. These days, cobalt-60 is used primarily
to sterilize medical equipment, such as surgical gloves. It has
been replaced in the treatment of cancer by X-ray beams or
electrons produced by linear accelerators.

In addition to identifying radioisotopes for use in medicine, the biology division also studied the effect of radiation
on living organisms. Nuclear reactors create large amounts
of radioactive elements, and the impact of radiation on
human beings became a growing concern during the war
years. Mitchell turned his attention to this matter, and began
a series of experiments exposing mice to radiation at the

Montreal Laboratory. He made several trips to the United States, where research in this field was more advanced, thanks to the Manhattan Project. The experiments he carried out in Montreal led Mitchell to reflect on the problem of how to precisely measure the doses of radiation received by the animals and, by extension, by humans. At first glance, it seems like a purely technical question, but in actual fact, it raises a fundamental point: If we don't correctly determine the doses, how can we know whether a worker, for example, has been exposed to dangerously high levels of radiation and, consequently, whether the protective measures in place are sufficient?

Defining the doses correctly meant that a single unit had to take into account the impact of the different types of radiation (alpha, beta, gamma, or neutrons), their intensity, and the way the person received the dose, e.g., external exposure, inhalation, or ingestion. Mitchell wrote a number of reports for the Montreal Laboratory on this subject, including Report HI-14: "Provisional calculation of the tolerance flux of thermal neutrons" and Report HI-17 entitled "Memorandum on some aspects of the biological action of radiations, with especial reference to tolerance problems." The tolerance he referred to was the ability of certain organisms to withstand high doses of radiation. The "tolerance problems" mentioned in the title referred to the challenges associated with determining the level of radiation mice—or humans—were able to withstand. After the war, the research undertaken by Mitchell and other scientists in the United States and Europe led to the development of dosimeters, which are small wearable devices used in radioactive environments to record the total radioactive dose absorbed by the user.

The work carried out at the Montreal Laboratory spurred significant advances in the field of human health. After all, the physicists who worked there had been handpicked from the world's top laboratories, and were at the cutting edge of atomic research. However, some of the countries that lagged behind in the frantic race to develop the atomic bomb weren't ready to throw in the towel; on the contrary, they would do whatever it took to catch up to the other competitors.

A DEN OF ESPIONAGE

The Gouzenko Affair

In the late afternoon on September 5, 1945, Igor Sergeyevich Gouzenko, a cipher clerk at the Soviet embassy in Ottawa, prepared to leave his office, knowing he would never set foot there again. He gathered his briefcase, which contained code books and about one hundred documents setting out detailed information about the Soviet Union's espionage network in Canada.

Encipherment is the transformation—using a code—of a clear message into an encrypted one that can only be read by someone possessing the key. It was a technique used by all the spy agencies. Gouzenko had decided to take a gamble and hand over top-secret Soviet information to the Canadians. He went to the RCMP detachment in Ottawa, but no-one there believed his story. So, he tried the *Ottawa Journal*, but the evening edition editor-in-chief wasn't interested either, suggesting he contact Canada's Ministry of Justice instead. Gouzenko headed to the Justice Department without delay, but the office was already closed for the day. Feeling more and more panicked, he rushed home to his apartment, where his wife and child were waiting. Gouzenko knew his life would be in danger if he was discovered, so the family hid in a neighbour's apartment for the night. Sure enough, a group of Soviet agents broke into the Gouzenko's apartment and rifled through their belongings. The agents were confronted by the Ottawa police and forced to leave, since they had no legal power to search a Canadian home. Igor Gouzenko had been in charge of ciphering in Ottawa since 1943. In September 1944, he had learned he was to be sent back to the USSR shortly. He feared, with good reason, that if he was being called back so abruptly, he must have committed some mistake which, in Stalin's Russia, surely meant he would face imprisonment or worse. Gouzenko was enchanted by the living conditions in Canada, and made the risky decision to defect. Over the course of the year, he photocopied top-secret documents, smuggling them out of the embassy and hiding them in his home. Some historians believe Gouzenko's defection that day—September 5, 1945— marked the beginning of the Cold War.

It would actually be more accurate to refer to September 6, because the following morning, Gouzenko went back to the RCMP office, where he finally managed to convince the officer in charge and, that very day, he met with the Minister of Justice, Louis St-Laurent. Gouzenko immediately requested political asylum. Saint-Laurent summoned Norman Robertson, Undersecretary of State for the Department of External Affairs who, in turn, informed Prime Minister Mackenzie King. For Canada's leader, it was a diplomatic time-bomb. Mackenzie King was reluctant to provide asylum to Gouzenko, so as not to upset the Soviet Union, which had been a military ally in the war that had just ended. Igor, his wife Svetlana, and their child were eventually taken to Camp X, a training camp for military commandos located near Oshawa, on the shores of Lake Ontario,

Igor Gouzenko preparing for an interview with the CBC. Following his defection in 1945, Gouzenko always wore a hood when making public appearances, for fear he would be attacked by the Soviet Secret Service. (Library and Archives Canada, MIKAN 3239912, fonds *Montreal Star*.)

a good distance from Ottawa. Gouzenko had numerous documents in his possession proving the existence of a vast network of spies in Canada that was targeting military-level scientific research, including, of course, atomic research.

Gouzenko had arrived in Ottawa in the summer of 1943, shortly after Colonel Nikolai Zabotin, a member of the GRU (the Russian Army's secret service) who, unbeknownst to the Soviet ambassador, was in charge of espionage in Canada. Zabotin had fought in the Battle of Stalingrad, in which the army of the German Reich suffered its first major defeat. The tall, distinguished colonel was well liked by the diplomats in Ottawa. Zabotin was very active for the GRU, and he

personally handled several spies in the network his predecessor in Ottawa had established. Moscow was especially interested in atomic research, but Zabotin had no-one working
within the Montreal Laboratory in 1943 and 1944. Zabotin
had complete faith in Gouzenko, allowing him surprisingly
free rein for an employee with access to top-secret documents. In violation of GRU rules, Gouzenko and his family
didn't live in the same building as Zabotin and the other
embassy staff. As the story goes, Zabotin's wife preferred
not to have to hear the Gouzenko's baby howl all night long
in the same building! Zabotin's mind was also elsewhere,
and he didn't keep an especially close eye on Gouzenko. He
had just begun an extramarital affair with Nina Farmer,
a Russian émigré who lived in Montreal, where she was
separated from her American husband. Zabotin, who often
travelled to Montreal on business, would stay at the lavish
Prince of Wales suite at the Ritz-Carlton, and would wine
and dine her there, and take her out dancing.[1] According to
Nina Farmer, the colonel even asked her if the name "Grant"
meant anything to her. She, of course, had no idea what he
was talking about, but it was an offhand question that could
have caused her considerable grief. Grant was the codename
Zabotin used in his secret communications with Moscow,
and the name his Canadian contacts knew him by.

In early 1945, Zabotin learned there was a sleeper agent
recruited by England who had been working for the Montreal
Laboratory for the past two years. When he found out, he
was furious with the Soviet military intelligence service,
which hadn't informed him of the agent's presence there,
apparently due to some GRU turf war.

He immediately jumped into action and sent GRU lieutenant Angelov to make contact with the spy in question

who had infiltrated the Laboratory. To his great surprise, the spy was neither German, nor French, nor even Canadian, but British!

Alan Nunn May

The GRU's contact at the Montreal Laboratory was a man named Alan Nunn May. To understand his transition from physicist to spy, one must delve a little into his past. Nunn May was born in 1911 in Birmingham, England, the youngest of four children. His father owned a brass foundry that was damaged by a fire at the end of World War I. The resulting economic hardship forced the family to move to a more modest home. Alan Nunn May went to school in Birmingham, and in 1929, he was accepted at the University of Cambridge, where he studied physics. During that time, he did a lot of reading on WWI, including Erich Maria Remarque's *All Quiet on the Western Front*, a novel that described the absurdity of trench warfare from the point of view of a young German soldier. From that moment on Nunn May became increasingly critical of the British establishment. In the 1930s, he was active in the Association of Scientific Workers, a trade union that represented employees of British universities from every level of the hierarchy. Nunn May took a keen interest in politics, and when Hitler was elected in Germany, he was deeply disappointed. The only country he felt was bringing a wind of change was the USSR. At about the same time, he became friends with Frederick Pateman, a fellow physics student from a working-class family who was a member of the British Communist Party. By the beginning of World War II, Nunn May felt increasingly at odds with the Party's position, which aligned with that of Stalin, who considered it an imperialist conflict on both sides. Like many British socialists and communists, Nunn May believed it more of an antifascist campaign, like the Spanish Civil War.

Although he was fascinated by all these ideas, Nunn May decided to focus instead on his physics studies. He was impressed by the work of Rutherford's group, and was desperate to join it. In 1939, he worked on radar in Suffolk, then in Bristol on a project to photograph high-speed particles emitted by radioactive elements. In 1942, he was approached by James Chadwick to join the Tube Alloys project. During the interview, Chadwick asked him if he had known a man named Nahum during his time at Cambridge (Ram Nahum was the most active communist among Cambridge's physics students). According to the memoir written by his adoptive son,[2] Nunn May replied:

> I said "Yes" and Chadwick went on: "We tried to get him [Nahum] for work on the project, but the security people made objections, on very silly grounds." And he looked at me with one of his quizzical stares, seeming to wait for me to get all the implications. The implications were very interesting. First, that Chadwick knew that like many of the nuclear physicists at the Cavendish I had been a party member, otherwise why raise the Nahum question at all? Second, that he personally regarded this as no bar to recruitment to the project, and so did the other scientists in charge. But the security people were liable to raise difficulties which made life awkward, so better not to ask any embarrassing questions, and would I please play it cool.

How extraordinary all the nuance that can be contained in a single glance!

He then met with George Thomson, the British physicist who won the Nobel Prize in 1937 and who chaired the

MAUD Committee (see chapter entitled "Laboratory at War," p. 49). Thomson complained that Halban's group had "too many damned foreigners," and had Nunn May sign the *Official Secrets Act* forms. And so, Alan Nunn May was recruited to join Halban and Kowarski's team at Cambridge. There, he worked on developing a more effective ionisation chamber. During that period, he received a visit from some Communist Party members who were researchers he had known from his days as a student at Cambridge. They urged him to renew his party membership and join a cell made up of scientists working on top-secret projects. Alan Nunn May never revealed the names of the people who had recruited him during his time in Cambridge in 1942, or the name of his cell leader.

In June 1942, Halban returned from a trip to the United States, where he had met with the people in charge of the Manhattan Project. While there, he had read a report on "radioactive poisons," or fission products that are created in an atomic pile and that could potentially be used in a conventional bomb dropped on a city, to render it uninhabitable. This is what is referred to today as a "dirty bomb." With these types of conventional bombs, i.e., that use a chemical explosive like TNT, radioactive material is added. This is different from an atomic bomb, in which the explosive is either uranium-235 or plutonium-239, whose lethal power can be thousands of times greater and can contaminate an entire region with radioactivity. Since the American intelligence service believed the Germans were on the verge of building an atomic pile, radioactive poison had become a priority.

Halban, who at that time headed up a team in Cambridge, tasked Alan Nunn May with compiling all the available information on radioactive poison and drafting a report setting out recommendations for the British authorities. Nunn May spent that summer working on the report, which he submitted on August 29. His findings were discussed at a meeting of the Tube Alloys supervisory committee in September 1942. The British and Americans feared the Germans would attack using radioactive poison. The Chicago Laboratory, where Fermi was developing the first atomic pile, was already preparing for a possible German attack on December 25, 1942. The scientists' families were moved out of Chicago, and Geiger counters were installed to detect any radioactive contamination from outside the Lab. As it turns out, their fears were unfounded, as the Germans trailed far behind the Allies in atomic research.

In his view, Nunn May felt that, if the Germans did in fact possess radioactive poisons, they were far more likely to use them against the Russians, especially since the key battles were taking place along the Eastern Front, not to mention the fact the Germans considered the Soviets an inferior species. That was what prompted Nunn May to take the first step that would lead to his undoing. During a conversation with his communist cell leader, he was asked whether information from the Tube Alloys project had been shared with the Russians. Nunn May replied that, insofar as he knew, nobody in Moscow was even aware of the project's existence. The cell leader told him he could arrange a meeting with a young Soviet diplomat who would be very interested in hearing what Nunn May had to say. So, Nunn May sat down and wrote a note summarizing what the English and Americans knew about radioactive poisons. The meeting between the two men took place at a London café at a table by the window. Just as Nunn May was passing his note to the diplomat, the Russian glanced out the window. Nunn May followed his gaze and spotted a man suddenly turn his back and walk away from the café. The British scientist had the impression it was a witness who had taken a photo that could be used against him as blackmail in the event he no longer wished to collaborate. Nunn May described this incident in a document that was found in his personal archives at the time of his death.

Alan Nunn May was part of the first group of scientists to arrive in Montreal. Before leaving for Canada, his cell leader told him to find himself an apartment away from prying eyes that could serve as a base for collecting nuclear secrets for the Russians. Nunn May protested: This wasn't what he'd signed up for. He was certainly willing to divulge important information, but he had no intention of becoming a professional spy! Despite his misgivings, he agreed to make contact with the Russians in Montreal. From there, things played out like a spy novel. Once he was settled in Montreal, Nunn May was to send a postcard to a young woman in London who was in on the subterfuge. Then, he was to expect a visit from someone who would approach him and say the words "Best regards from Mikel" ("Mikel" being the codename of his mentor in Canada). With all this cloak-and-dagger intrigue, Nunn May must have had a lot on his mind when he arrived in Montreal.

Just like for the other Montreal Laboratory scientists, 1943 was a difficult year for Nunn May due to the lack of collaboration from the Americans. One of the things he was working on was the absorption of neutrons by a rare isotope of oxygen, O_{17} (oxygen-17), which accounts for 0.04 percent of the composition of natural oxygen. Oxygen-17 absorbs neutrons much more readily than its more abundant cousin, oxygen-16. Nunn May came up with an experiment to measure the absorption of oxygen-17, something that had never been done before. His work provided valuable information to the Montreal Laboratory in its quest to build a heavy-water reactor. If the oxygen-17 was too absorbent, its reactivity would need to be boosted, for example, by further enriching it with uranium-235, to achieve pile criticality. Nunn May set up his experiment in a separate room. Since the neutron source he was using was weak, he had to avoid any electromagnetic interference, to ensure his ionization chambers (used as detectors) would function properly. He covered all the room's walls with a paper coated with a thin layer of copper, which not only acted as an electromagnetic screen, but also lent it a rather exotic, opulent air. Nunn May's experiment worked, and he was able to obtain preliminary values on the absorption of neutrons by oxygen-17.

When John Cockcroft was appointed director in early 1944, visits between the Montreal and Chicago laboratories were once again allowed. Approval was granted to Alan Nunn May to replicate his experiment at the Chicago Lab, which had access to much more powerful neutron sources than Montreal did. As a result, he was one of the first "Montreal" scientists to visit Fermi's group, where he worked in collaboration with an American physicist named Herbert Anderson. With the help of his assistant Ted Hincks, a young Canadian physicist, Nunn May had all the necessary equipment shipped to Chicago. Before his first trip there, Nunn May's engineer and physicist colleagues in Montreal gave him a list of questions, the answers to which would be very useful in advancing the heavy-water atomic pile project. Until then, the Canadians hadn't been able to acquire such information without going through official channels, which was a slow process. When he arrived in Chicago,

Nunn May explained to Herbert Anderson his Montreal colleagues' concerns, and showed him their list of questions. Anderson agreed with Nunn May that the restrictions imposed by General Groves and the other security officials were too stringent and unnecessarily taken to the letter. He subsequently provided the British scientist with a large number of reports, which Nunn May would read during his stays in Chicago, taking notes as he went and showing them to his colleagues back in Montreal, much to their delight, all under the noses of the security officials. Alan Nunn May repeated his experiment as planned, and was able to confirm that the absorption of oxygen-17 was low enough that it wouldn't have an impact on the Canadian reactor. With that, he became a regular visitor to Chicago. On his final trip to the Windy City, he learned from Anderson that a crucial experiment on the properties of uranium-233 and uranium-235 had just been conducted. Before he left, Anderson showed him the samples of the two uranium isotopes, divided them in two, and gave half to Nunn May, urging him to replicate the experiment in Montreal. This is very surprising and hard to believe, given how few samples of these substances existed, particularly uranium-233, which is not found in nature. However, during his trial, Nunn May insisted that the exchange had taken place.

One day in early 1945, Nunn May received a phone call at the Laboratory from a man with a strong foreign accent who wanted to meet with him to pass on a message. Intrigued, Nunn May agreed. The first thing the man said when they met was, "Best regards from Mikel." As the English physicist would later learn, his contact was Pavel Angelov from the Soviet Embassy in Ottawa who'd been sent by none other than Nikolai Zabotin. Angelov explained Nunn May's mission to him: he was to compile as much information as possible on the pile under construction in Chalk River. The scientist gave it a moment of thought, then made up his mind. Although by early 1945, the German defeat was only a matter of time, and the Russians were no longer under threat, he decided to collaborate because he wanted to prevent the United States from being the only nuclear power in the post-war age. He knew that the British were also in the process of developing, independently of the United States, atomic piles and probably atomic bombs. It was also widely believed that the French had the same ambition.

Angelov and Nunn May worked out a method for passing along information. At home in his apartment, Nunn May would type out reports summarizing what he knew about the work being done in Montreal and Chicago. For the Tube Alloys official reports, it was more complicated, since they weren't supposed to be removed from the Laboratory. However, in reality, the lead researchers would often take these reports home to read them over the weekends. So, Nunn May would borrow the reports from the Montreal Laboratory library on Friday, give them to Angelov that evening, and Angelov would copy them out on Saturday and bring them back to Nunn May on Sunday. In this way, Nunn May divulged to the Soviets the main research being conducted on the chemistry and metallurgy of uranium and plutonium, as well as the problems that had been encountered and the solutions proposed in designing the heavy-water pile. The reports were then encrypted and sent to Moscow by Igor Gouzenko.

When Anderson gave Nunn May the samples of uranium-233 and uranium-235 in Chicago, instead of taking them back to the Montreal Laboratory to take new readings,

Nunn May handed them over to Angelov, who shipped them to Moscow. The samples contained only a few milligrams, and were easy enough to conceal. Nunn May continued his subterfuge right up until the summer of 1945, keeping Moscow informed of all the rumours flying at the Montreal Laboratory about the Americans' atomic testing in New Mexico. After Hiroshima and Nagasaki, the Americans published the Smyth Report, which outlined the atomic research that was carried out during World War II, with the exception, of course, of the details about building the bombs. When the ZEEP nuclear reactor in Chalk River first went critical on September 5, Nunn May's presence was no longer required in Montreal and, moreover, the British were about to start up their research at the Harwell Laboratory. When Alan Nunn May boarded a plane and set off for London in September 1945, he was confident his career prospects were very bright indeed.

Yet, throughout the fall and early winter of 1945, Nunn May found no work at all. Gouzenko's defection had shaken things up and, although Nunn May didn't know it, he was being closely watched. On February 15, 1946, Michael Perrin, the Tube Alloys scientific coordinator for the British government, summoned Nunn May to his office in London. Nunn May thought he was being called in to discuss the role he would play in Britain's atomic research project. Instead, Perrin introduced him to two men, Leonard Burt and Reginald Spooner, who worked for MI5, Britain's domestic counter-intelligence and security agency. The first words out of Burt's mouth were "Best regards from Mikel." He immediately accused Nunn May of spying for the Russians, which Nunn May vigorously denied. That same day, thirteen Canadians were arrested in the Gouzenko Affair. In their

subsequent interrogations, the MI5 agents made it clear
to Nunn May that they had irrefutable proof against him
and that he would, in all certainty, be convicted. If Nunn
May continued to deny everything, he could be extradited
to Canada and, from there, possibly to the United States,
where his sentence would likely be much harsher than in
England. Nunn May was arrested on March 6, and his trial
got underway on May 1. The trial attracted widespread
attention, making headlines in Canada and Britain alike.
Nunn May pleaded guilty to violating the *Official Secrets Act*.
His lawyer, Gerald Gardiner, asked for a reduced sentence,
arguing that the Soviet Union was an ally at the time of the
events. But Justice Oliver rejected the plea and sentenced
Nunn May to ten years in prison, although he was released
early for good behaviour, on December 29, 1952. When they
saw the mob of journalists waiting at the prison gates, the
authorities allowed Nunn May to leave through a side door,
and from there he was escorted to the train station, where
he boarded a train to go and stay with his brother.

The media tracked him down all the same and camped
out in front of his brother's home until Nunn May finally
issued a written statement saying that he would give no
interviews for the rest of his life and that he rejected the
accusation of treason, claiming it was untrue because the
only thing he had wanted was to see the Allies win the war
against Nazi Germany and Japan, and that from that day on,
he planned to devote himself to the betterment of society
through scientific research.

Fulfilling this last objective would turn out to be very
challenging for Nunn May, as he had become persona non
grata, and nobody would hire him. He later met Hilde
Broda, the ex-wife of Engelbert Broda, an Austrian physicist

Physicist Alan Nunn May (seen here in 1953) passed on confidential information, and even radioactive products, to the USSR while he was working at the Montreal Laboratory. (International Press, personal archives of Paul Broda.)

who had sought refuge in England, and whom Nunn May had worked with at the Cavendish Laboratory just prior to leaving for Montreal. Hilde and Alan Nunn May were married in 1953, and Alan became the adoptive father of Paul, the son of Hilde and Engelbert. Their marriage gave rise to much speculation, as Engelbert Broda was a member of the Communist Party, and Hilde was a party sympathizer. Both men always categorically denied any connection between Alan's espionage activities and the Brodas. Despite Nunn May's employment woes, Alan and Hilde were happy together. In 1961, Ghanaian president Kwame Nkrumah invited Nunn May to teach physics at the University of

Ghana, in Accra. He accepted the position, and Hilde joined him there in 1962. He was later appointed head of the physics department, where he conducted research in solid-state physics. He also created a science museum there. Alan Nunn May retired in 1978 and returned to live in Cambridge, where he died January 12, 2003, at the age of 91. While his case garnered widespread attention, Alan Nunn May was by no means the only scientist spy during World War II.

Britain and the United States suspected that the Russians, even though they were Allies, were spying on them under their noses. General Groves, in particular, was somewhat paranoid about the prospect. He had a visceral hatred of communists, and harboured suspicions about anyone who wasn't an American, in fact, anyone who had fled Europe seeking refuge, especially the French. Despite his enormous responsibilities, he would still make a point of monitoring the comings and goings of the Montreal Laboratory scientists and when, in the spring of 1945, his suspicions were aroused by Nunn May's frequent visits to Chicago, he brought the matter to Director Cockcroft's attention. Cockcroft vouched for Nunn May, as Chadwick had done before him when the physicist was hired. After all, Nunn May was a bona fide Brit who had studied at Cambridge. And as far as the British were concerned, so long as you were from the right social class, you were considered above suspicion. If there was one person Groves particularly disliked, it was Halban, likely because of his diverse background (Austrian and French) and the fact he spoke several languages, and this was no doubt part of the reason Groves had him replaced by Cockcroft as director of the Laboratory. After Halban's regrettable journey to Paris in December 1944 to meet with Joliot, Groves and the Americans essentially blacklisted

Halban, preventing him—and the other French scientists, including Goldschmidt—from returning to France right away when the war ended.

The British Secret Service had even begun enquiries into Fritz Paneth, despite him having left his homeland and devoted many years of his life to the Allied cause fighting against Hitler. When Paneth worked for the Tube Alloys project in England in 1942, his neighbours had denounced him to the police as they suspected he was a spy because he spoke German with his wife at home and they never rose before 10 in the morning![3]

However, there was one man who managed to slip through the fingers of Groves and both the British and U.S. Secret Services, a man who, by all accounts, was also passing on information to the Russians. He was active within a rival organization of the GRU, the state security service known as NKVD, which was the predecessor to the KGB. The mysterious story of this key figure at the Montreal Laboratory sheds additional light on the leaks about Montreal's nuclear secret, leaks that continued even after the war ended. His case is crucial to grasping the full extent of the details that were secretly divulged about the Canadian laboratory.

Bruno Pontecorvo, A Central Figure

As we saw earlier, Enrico Fermi was one of the most brilliant nuclear researchers, and the success of the American project was largely driven by his intelligence and charisma. He also trained numerous researchers in his methods, and getting its hands on one of these top-tier scientists would be a real coup for Montreal's cutting-edge nuclear project. In November 1942, at the recommendation of Bertrand

Goldschmidt, Halban and Placzek met with one of these physicists, a man named Bruno Pontecorvo. The Montreal Laboratory had just been founded, and the timing was perfect for Pontecorvo, who had been growing increasingly frustrated working at an oil exploration company in Oklahoma. Halban and Placzek explained to Pontecorvo the role they had in mind for him at the Montreal Laboratory: he was to be involved in designing a heavy-water atomic pile. The pile would provide energy in the post-war period, as well as the fissile isotopes of plutonium that were needed to build an atomic bomb. Pontecorvo accepted the offer, and his family began to prepare for the move. He resigned from his position in Oklahoma in early January 1943 and headed to New York, where he was officially hired by the Tube Alloys project as a member of the experimental physics group under the leadership of Frenchman Pierre Auger. He stayed in New York for a few weeks, where he was brought up to speed on the latest developments in atomic research. Then, on February 7, 1943, Bruno, his wife Marianne, and their son Gil arrived in Montreal. After the balmy weather of Oklahoma, Montreal in winter was something of a shock, with temperatures regularly dropping below minus 20°C. The couple rented an apartment near Chemin de la Côte-des-Neiges that overlooked St. Joseph's Oratory and was within easy walking distance of the Laboratory at the new Université de Montréal building.

Pontecorvo would be involved in most of the projects underway at the Montreal Laboratory. While he was specifically assigned to the experimental physics group, he would often mingle with the chemists and theoretical physicists, and would engage in lunchtime discussions with them about theoretical physics, but also about philosophy and current events. At about the same time, he began to take an interest in neutrinos—a subject that would occupy most of his scientific life after the war. Once the situation with the Americans eased in 1944, Pontecorvo travelled frequently to Chicago, and he was the one who signed most of the reports. The British supervisors deeply valued his contribution, and the expertise he had acquired searching for oil deposits with Well Surveys in Oklahoma came in handy, this time to locate new uranium deposits in Canada. Pontecorvo travelled to New York to meet with Gilbert LaBine, the president of Eldorado Mines, who owned several large uranium mines in the Northwest Territories. Pontecorvo even spent several days in September 1944 in Port Radium, near Great Slave Lake, to help regulate the company's prospecting instruments. He also worked on the ZEEP reactor measurement instruments and on the design of a reflector, the mass of water around the outside of the pile that, as the name suggests, serves to reflect the neutrons that might otherwise escape back towards the core of the pile. The Italian scientist was present in Chalk River on September 5, 1945, when the ZEEP reactor (the little reactor Kowarski had "sired") went critical. In other words, Pontecorvo was closely involved in the activities of the Laboratory. Alma Chackett has fond memories of the man and of his "very pretty wife," who got along well with the young employees. But what nobody at the Montreal Laboratory knew was that there was a dark shadow lingering in Bruno Pontecorvo's past.

Who Was Bruno Pontecorvo?

Bruno Pontecorvo was born in August 1913 in Marina di Pisa, in the Tuscany region of Italy. He was the fourth of eight children in a tight-knit Jewish family. His four brothers and three sisters all went on to have illustrious careers: The eldest, Guido, was a renowned geneticist in Great Britain; Paolo became an engineer; Gillo, with whom Bruno was especially close, became a film maker whose most famous film is *The Battle of Algiers*; and his sister Giuliana was a journalist. Their father came from a well-off family in Pisa, and owned a textile factory. Pontecorvo Senior was above all a committed anti-fascist and was on close terms with his workers.

From a very young age, Bruno showed an aptitude for math and science. At the urging of his brother Guido, he went to Rome in 1931, at the tender age of 18, to join Enrico Fermi's group. Guido Pontecorvo was good friends with Franco Rasetti, a physicist and close friend of Fermi's. Guido and Franco both belonged to the same mountaineering club, and had climbed several mountains together. After passing the perfunctory exam Fermi gave him, Bruno enrolled at the University of Rome, where he completed his degree in physics and officially joined Fermi's team in 1934, a year of great discoveries. Pontecorvo co-authored two of the Rome group's most important articles on inducing artificial radio-activity in uranium using water and paraffin wax, which slowed the neutrons. Pontecorvo was also included on the patent application filed by Fermi regarding this discovery. In 1936, as the fascists became increasingly anti-Semitic, Pontecorvo fled Italy to join the Curie Laboratory in Paris, where he continued his research for another four years. During his stay in Paris, where he resided at the grandly named, albeit poorly appointed, Hôtel des Grands Hommes, Pontecorvo made the acquaintance of Marianne Nordblom, a young Swedish woman who had travelled to Paris to learn

French. They struck up a relationship that was to last their entire lives. In her book about the life of her husband, Enrico, Laura Fermi writes that Bruno was uncommonly good-looking. He fit the stereotypical image of the charming Italian who likes to be the centre of attention at parties.[4] Marianne got pregnant and moved in with Pontecorvo in early 1938, although they were not married—a highly unusual situation for a young woman in those days. Marianne gave birth to a boy, Gil, in July. The following two years were difficult for the couple: their financial situation was dire, and Marianne's visa had expired. A mere six weeks after the birth of their son, she was forced to return to Sweden, leaving Gil in the care of a nursery north of Paris. She was finally able to return to Paris more than a year later, in early September 1939, just as war broke out. Bruno and Marianne were married in Paris at the beginning of 1940.

Paris was also where Bruno Pontecorvo, encouraged by several members of his family, including his cousin Emilio Sereni, began to develop an interest in communism. His personal correspondence reveals that Pontecorvo was very concerned about the growth of fascism in Europe. In late August 1939, he became a member of the French Communist Party, a decision that would have major consequences later in his life.

When the Germans invaded France, Pontecorvo managed to secure a travel permit for Marianne and Gil for a one-way journey to Toulouse, where his sister Giuliana and her husband had been living for the past two years. Marianne and Gil boarded a train in Paris with all their luggage on June 3, and arrived safe and sound at Giuliana's home. Bruno remained in Paris until the Germans began closing in on the city. Frédéric Joliot wanted him to accompany Kowarski and Halban to Clermont-Ferrand, but French intelligence were against the idea, ostensibly because Bruno wasn't a French national, although his political convictions may have had something to do with the decision. Several members of the Pontecorvo family were still in Paris when the German army began marching on the city. The day before the German occupation began, the family decided to flee, escaping just in the nick of time. There was no transportation available, and with hundreds of thousands of people clogging the roads heading south out of the city, the Pontecorvos took to their bicycles. Their plan was to ride until they could find a taxi or a train that would take them the rest of the way. Pedalling their heavily laden bikes, they managed to cover sixty kilometres the first day, stopping to spend the night at an inn. They woke the following morning to the sight of the village square teeming with German tanks. They decided to push on, weaving their way around the broken-down vehicles littering the road. Fortunately, they weren't stopped, and they made it to Orléans. With so many people on the move, finding a room for the night was out of the question, and many simply slept wherever they could find a patch of grass.

After ten days of cycling, the Pontecorvos finally reached Toulouse, where Bruno was reunited with his wife and son. By a stroke of pure luck, another former member of the Rome group, who was by then already in the United States, had sent Bruno a job offer for an oil prospecting company in Oklahoma. With that letter in hand, Pontecorvo began the process, in the midst of the war, of emigrating to America. The Pontecorvos left Toulouse by train on July 19, and five long days later, finally arrived in Lisbon. After the bedlam of the previous two months, they were surprised at how calm things were there. The war seemed like a distant memory. Sadly, Marianne, who was three months pregnant at the time, suffered a miscarriage. Although she didn't feel up to the trans-Atlantic voyage, they had no choice but to go, and on August 19, 1940, their ship docked in New York City. From there, as we know, Pontecorvo worked in Oklahoma until he was hired by the Montreal Laboratory in early 1943. This phase of his life culminated in the commissioning of the ZEEP reactor on September 5, 1945.

The Sudden Disappearance of Pontecorvo

After the war, once the little nuclear reactor went critical, Bruno Pontecorvo received a flurry of attractive offers from American and European universities alike. At first, to the surprise of his colleagues, he decided to join the team at the Harwell Laboratory in England, then changed his mind and stayed on in Chalk River. He explained that he had worked hard on the NRX, the heavy-water reactor being built in Chalk River, and he wanted to see it through to fruition. He was, in fact, one of the only physicists present in the control room when the NRX became operational on July 21, 1947. At the time, the NRX was the only reactor in the world with such a high flux of neutrons, making it the ideal tool for research. During his stay in Chalk River, from 1945 to 1947, Pontecorvo was increasingly focused on his research into neutrinos. In 1946, he penned a new report in which he mentions for the first time the solar neutrino problem, a notion that would make him famous. Nuclear reactions that occur in the Sun emit astronomical amounts of neutrinos. The flux of solar neutrinos on the Earth's surface is some ten billion per square centimetre per second. Pontecorvo proposed a method to detect neutrinos that would eventually come to be accepted several years later.

In 1949, Bruno Pontecorvo accepted a position at the Harwell Laboratory in England, where he worked on the British nuclear energy program. He also conducted basic research on cosmic rays. He and Marianne settled in Abingdon-on-Thames, a village halfway between Harwell and Oxford, with their three sons Gil, Antonio, and Tito. In the summer of 1950, the family went on a camping trip together with Bruno's sister Anna, to France, Switzerland, and Italy. They left England on July 24, arrived at Lake Como on July 31, and planned to return on September 7. Anna left the Pontecorvos for a few days, while Bruno, Marianne, and the children continued on to the Dolomites. From there, they would go to visit Bruno's parents in Milan. On leaving Milan, he arranged to meet up with his parents again in Chamonix at the end of August, since he would be going there to conduct an experiment on cosmic rays. The family took to the road again, heading south to Ladispoli, near Rome, where Giuliana, another of Bruno's sisters, was spending the summer. Bruno and Marianne then continued on to Circeo, a seaside resort a couple of hours south of there, where they pitched their tents. On August 22, Bruno celebrated his 37th birthday in the company of his brother Gillo and sister Anna. Up until then, nothing seemed out of the ordinary. But, on August 24, the day he was supposed to meet up with his parents in Chamonix, Bruno didn't show up. Instead, he sent them a telegram explaining he'd had car trouble. Then, on August 29, he and his family abruptly disappeared without a trace. He didn't return to Harwell in early September and, in fact, for the next five years, there was no news whatsoever of his whereabouts, which, unsurprisingly, gave rise to all kinds of rumours.

In February 1955, the cloud of mystery hanging over Pontecorvo's disappearance suddenly lifted when an article appeared in the Soviet newspaper *Izvestia*, in which Bruno Pontecorvo was interviewed, explaining that he had defected to the USSR because of the "moral suffering" he had experienced as a physicist following the bombing of Hiroshima and Nagasaki, and because of the heavy-handed police interrogations he was forced to endure when he worked at Harwell. The events that led up to Pontecorvo's defection in the middle of his summer holidays are admirably recounted by Frank Close in his biography *Half-Life, the Divided Life of Bruno Pontecorvo, Physicist or Spy.*[5] Here is how Close pieced the story together:

As we know, Pontecorvo joined the French Communist Party while he was working at the Joliot Laboratory in the early days of the war. He maintained his convictions throughout the war, although he was very careful to keep them to himself once he arrived in the United States, and when he worked in Montreal and Chalk River, and at the Harwell Laboratory. In fact, many of his colleagues assumed he was entirely apolitical. In the fall of 1942, while he was in discussions to join the Montreal Laboratory team, the FBI searched his home in Tulsa, Oklahoma, and questioned his wife. The federal agents drafted a report in which they described Pontecorvo as a communist sympathizer and suggested he not be given a position where he would be involved in any top-secret military projects. The report was mentioned by the FBI in an exchange of correspondence with British security officials in Washington, but due to an administrative error, the letter was never passed on to the appropriate authorities, and Pontecorvo received his security clearance to work on the Tube Alloys project. The report in question sat unread for nearly eight years.

Then, towards the end of 1949, the hysteria sparked by McCarthyism led to the employees of the U.S. government and universities being required to pledge a "loyalty oath," stating whether they harboured any communist sympathies or knew anyone who did. Emilio Segrè (photographer of the cover-page photo), one of Pontecorvo's colleagues in the Rome group, worked at the University of California, which was particularly zealous about obeying Senator McCarthy's orders. Segrè met with Robert Thornton, a friend who worked at the newly formed Atomic Energy Commission (AEC), and told him that he was aware that several of Pontecorvo's brothers and sisters were openly communist, and that his cousin, Emilio Sereni, was a communist and member of the Italian government. Thornton passed the information on to the FBI, which went back to its Pontecorvo file, and discovered the 1942 report on him. The FBI informed its British counterpart, MI5, but the British Secret Service took several months to react, probably because they were preoccupied with the handling of the Klaus Fuchs case in the early days of 1950. Fuchs, a German-British physicist and committed anti-fascist, worked at Los Alamos where, as it turned out, he was passing along top-secret information to the Soviets. It was Fuchs who divulged critical information about the mechanisms of the atomic bomb to Moscow. He was found out thanks to an American decryption program that also incriminated Ethel and Julius Rosenberg, who were executed by electrocution at Sing Sing in 1953.

By the spring of 1950, Pontecorvo sensed that he had aroused the suspicions of the authorities. In late April, he travelled to Paris to attend Frédéric Joliot's 50th birthday party. The atmosphere was celebratory and festive, that is, until Joliot got up to announce that, that very same afternoon, he had been relieved of his position as director of France's Atomic Energy Commission because he was openly communist. He continued to address the partygoers for what seemed like hours, and Pontecorvo came away shaken to the core. Over the course of the summer, a British secret agent who worked at Harwell, Henry Arnold, had several discussions with Pontecorvo during which the Italian scientist admitted that, while several members of his family were communists, he himself was not. At the insistence of MI5, which wanted to ensure he was removed from any further atomic research, Pontecorvo was offered a position as a professor at Liverpool, where a large particle accelerator was under construction. Pontecorvo had seen what was happening in the U.S., with McCarthy's witch-hunt, and this was most certainly on his mind when he accepted the job in Liverpool and left on holiday with his family. But his fate was already sealed, as a series of events occurred during the summer.

On July 13, the MI5 representative in Washington, Geoffrey Patterson, wrote to his superiors in London about Bruno Pontecorvo, informing them that the FBI had information dating back to 1942 revealing that Pontecorvo and his wife were communists. It appears that the American Secret Service were looking to further their investigation into the Pontecorvos. By all accounts, a Soviet double agent who was working for MI6 in Washington intercepted Patterson's message about the Pontecorvo couple and passed it on to his contacts in the USSR. With Klaus Fuchs and most of the spy ring in the United States now exposed, the Soviets were keen to limit the damage and get their as-yet-unidentified spies out of the country. Frank Close, in his book *Half-Life, the Divided Life of Bruno Pontecorvo, Physicist or Spy*, hypothesizes that

Internationally renowned physicist Bruno Pontecorvo, one of the key figures at the Montreal Laboratory, was suspected of being part of an espionage ring that sought to help the Russians catch up to the Americans in the field of nuclear science. These 1944 photos were taken during lunch break in Montreal. Above Pontecorvo is walking with Pierre Auger. On the second photo he is walking with Grace McKercher and Elizabeth O'Brien, secretaries of the Physics group. (Personal archives of Grace Stewart, courtesy of the Society for the Preservation of Canadian Nuclear Heritage Inc.)

the Soviets threatened to reveal Pontecorvo's espionage activities in order to force him to defect.[6]

After travelling to Rome with his family, Bruno Pontecorvo booked a flight for them to Stockholm. On September 1, without a word to anyone, the family flew to Stockholm, where they spent the night at a hotel, courtesy of the Soviet Embassy. The next morning, the Pontecorvos boarded another flight, this time for Helsinki. Two vehicles came to pick them up on the morning of September 3. Marianne and the children piled into one, while Bruno hid in the trunk of the other. The convoy made its way through vast stretches of Finnish forest, crossing the border into the USSR at Vyborg. After another 100 kilometres or so, the cars pulled over and Bruno climbed out of the trunk. With that, at the age of 37, he left behind his work, his friends, and his brothers and sisters. And so began the final episode of his life, a life that would end in the Soviet Union.

This hasty escape suggests that Pontecorvo was spying for the Russians and that his involvement wasn't brought to light by the Gouzenko Affair. The spies who were exposed (as well as the Cambridge Five and Klaus Fuchs) were all working for GRU (Soviet military intelligence). Another ring organized by the KGB existed parallel to the GRU network. The KGB was responsible for state security, the political police, and intelligence. GRU was subordinate to the military, while the KGB answered directly to Stalin. We know that Stalin insisted that intelligence from one agency be corroborated by agents from the other before he would trust it. If Bruno Pontecorvo was passing on information via KGB agents, Gouzenko would not have known about it, which is perhaps why Pontecorvo's name wasn't mentioned in the documents he handed over to the Canadian authorities.

At any rate, Pontecorvo and his wife spent the rest of their days in the USSR. Bruno was assigned to the nuclear research centre that had just been built in Dubna, a town about a hundred kilometres north of Moscow. The centre was home to the world's largest particle accelerator, a claim it held until the Brookhaven accelerator in New York went into service three years later. Pontecorvo turned his attention back to his research on neutrinos, a field he continued to devote himself to for the rest of his days. He put forth a number of fundamental ideas about neutrinos and their anti-matter particle—antineutrinos—and argued, in 1959, that there were actually two different neutrinos: electron neutrinos (v_e) and muon neutrinos (v_μ). When solar neutrinos were first detected in experiments, there was a problem: only one-third of the predicted number were found. That is when Pontecorvo proposed a solution dubbed "neutrino oscillation," suggesting that neutrinos "oscillate" between the different types, and that one-third of the solar neutrinos transform into muon neutrinos while another third transform into tau neutrinos (v_τ) during their voyage between the Sun and the Earth. After his defection to the USSR, Pontecorvo never returned to Canada. But in an ironic twist, it was actually a laboratory in Ontario that confirmed his theory. In 2015, Art McDonald, the director of the experiments conducted at that Canadian lab, received the Nobel Prize in Physics. One can only speculate that if Pontecorvo hadn't defected in 1950, he would have been awarded the Nobel many years earlier.

During his first five years in the Soviet Union, the authorities prohibited Pontecorvo from having any contact with the outside world. Starting in 1955, and for the next 23 years, the only contact he was allowed was to mail letters to his family. It wasn't until 1978 that he was given permission to travel outside the USSR and was finally able to reunite with his sister Anna after so many years apart. That same year, he began to develop symptoms of Parkinson's, and the disease would eventually kill him. The final years of Pontecorvo's life were rather sad. After the collapse of the Soviet Union, he gave a long interview to Miriam Mafei, an Italian journalist and communist, in which he said he regretted having to endorse the Soviet authorities' decisions all his life. He continued to work at Dubna until his death on September 24, 1993, at the age of 80.

As we have seen, Gouzenko had no information to share about Pontecorvo's possible role as a spy. He did, however, have plenty of secrets to spill about dozens of Canadian scientists.

The Kellock-Taschereau Commission

In the wake of Gouzenko's defection, the Canadian authorities contacted the British and American governments. Together, they decided to keep his revelations under wraps, at least until they had thoroughly investigated all the individuals he had named. Thirteen Canadians who were implicated were arrested on February 15, 1946, and twenty-six more on March 14. Of the thirty-nine, eighteen were found guilty of espionage, sixteen were acquitted, and five were released without charges. "The Gouzenko Affair," as it was quickly dubbed by the media, forced the Canadian government to launch a royal commission to investigate espionage in Canada—the Kellock-Taschereau Commission—which it did on February 5, 1946. Among the accused were Fred Rose, a Montreal MP and Communist Party member; and Sam Carr, one of the leading organizers for the Communist

Party of Canada. Both men were found guilty and sentenced to ten and seven years' imprisonment, respectively. After serving their sentences, they emigrated to Poland, where they lived out their days. Gouzenko's most sensitive revelation concerned the Montreal Laboratory. He claimed that two scientists had passed on documents, plans, and even radioactive material to the Soviet Embassy and to contacts in the United States. In addition to Alan Nunn May, Gouzenko implicated Norman Veall, a young physicist who worked in Nunn May's department. At his trial, Veall was completely exonerated by his colleague and was never charged. After the war, Veall went on to work in nuclear medicine at several London hospitals, and become a world expert in the field.

In the spring of 1946, the Kellock-Taschereau Commission implicated nine scientists, including Alan Nunn May and Norman Veall, in espionage activities for the Soviet Union. Among the others were Raymond Boyer (chemist, McGill University, RDX Project), Israel Halperin (mathematician, Canadian Armament Research and Development Establishment), and Phillip Dunford Smith (physicist, NRC, radar research).[7] Halperin, who refused to testify at his trial, was eventually acquitted. Boyer was sentenced to two years in prison, and Dunford Smith to five years.

On closer examination of the spies who worked on behalf of the Soviet Union from within several key Canadian, British, and American military projects, it appears that Nunn May's and Boyer's contribution was relatively limited. The most valuable information by far came directly from Los Alamos, where the Americans developed the theory and precise mechanisms behind the uranium and plutonium bombs. Nevertheless, that doesn't diminish the fact that, as the first proven instance of Soviet espionage in

Allied territory, the Gouzenko Affair was, in many ways, a catalyst for the Cold War. When the Smyth Report was released on August 12, 1945, the "secrets" revealed by Alan Nunn May were nearly all made public. Needless to say, even without the intelligence it gathered through its network of spies, Russia's scientific knowledge was advanced enough to allow the country to build the bomb. That said, it is estimated that, with the information gleaned from Los Alamos and other nuclear sites, Soviet scientists and engineers were able to expedite their development of the A-bomb by two full years. In other words, without the help of its spies, the USSR wouldn't have exploded its first atomic bomb, nicknamed "Joe One" by the Americans, until 1951, instead of in 1949 as it did. After Hiroshima and Nagasaki, Stalin became obsessed with developing his own bombs to counterbalance the Americans. The twin nuclear bombings of Japan on August 6 and August 9, 1945, changed everything, both in terms of international politics and on an individual level for millions of people, including the employees of the Montreal Laboratory.

CONCLUSION

Hiroshima and Nagasaki

The Montreal Laboratory employees learned about an atomic bomb having been dropped on Hiroshima at the same time as the rest of the world. Chemist Alma Chackett recalls hearing it from Leo Yaffe, who had gone downtown on the morning of August 7 and had spotted the newspaper headlines. He brought a copy of the paper back to the Lab, where the news was met with somber silence. Some of the people at the Montreal Laboratory were aware of one of the goals of their research, namely to produce material to build a new and extraordinarily destructive weapon. About twenty of the Montreal Laboratory employees knew that the Americans had managed to build a bomb in June 1945 and that a test explosion had been conducted. But most of the employees had only a very vague idea of the true objective of the Lab's research, and were stunned to learn that their project was actually connected to the atomic bomb.

Four days after the bombing of Nagasaki, the Canadian government issued a statement providing details about the Tube Alloys project and purpose, and about the Montreal Laboratory, along with a list of the scientists who worked there. It is hard to imagine today a government press release going into such great detail about a matter of basic and applied science. The document was entitled "Canada's Role in Atomic Bomb Drama," and its second paragraph reads: "The dropping of the first atomic bomb is [...] the culmination of the work of scientists from many nations, the pooling of the scientific and natural resources of the United States, Britain, and Canada, and the expenditure of hundreds of millions of dollars in the United States and smaller, but substantial, sums in Canada on plants and equipment in the most extensive scientific effort ever directed towards the attainment of a new weapon." It goes on to briefly describe the Montreal Laboratory and the Chalk River site. One question that arises frequently in relation to the Montreal Laboratory is whether Canada was directly or indirectly involved in the dropping of the nuclear bombs on Hiroshima and Nagasaki.

At the beginning of the war, when the Tube Alloys project was established, the objective was clear: to develop an atomic bomb before the Germans did. The idea of an A-bomb in the hands of Hitler was a terrifying prospect. Given the crimes against humanity they had already committed, the Germans would likely have had few qualms about using the weapon against England and other countries. Developing a bomb was therefore considered ethically justifiable for the scientists working on Britain's atomic project, especially those who had fled the Nazi invasion: Frisch, Peierls, Paneth, Kowarski, Halban, Pontecorvo, Goldschmidt, and others. And like with all megaprojects, a dynamic emerged in which achieving the final goal became the primary shared objective of those involved. Subsequently, building the bomb appeared as a means to bring an end to the conflict. This was clearly how Groves, Oppenheimer, and Roosevelt viewed the

Manhattan Project. They knew they would have to answer to Congress once the war was over. How else could they justify the project's cost of nearly $2 billion?

But by May 1945, faced with the imminent production of the atomic bomb, some of the Manhattan Project scientists began to have doubts. In June, at the initiative of James Franck, a group of seven scientists at the Chicago Laboratory wrote a document[1] proposing that nuclear weapons be brought under international control so as to avoid an arms race, and suggested that a demonstration explosion be conducted rather than using the weapons against Japan. The note, known as the Franck Report, was made public in 1946.[2] A meeting of the interim committee overseeing the Quebec Agreement between the British and the Americans was called to consider James Franck's report and make a recommendation to military and political leaders. The committee, in turn, consulted its scientific panel, which consisted of Arthur Compton, Enrico Fermi, Ernest Lawrence, and Robert Oppenheimer. After much debate, the scientific panel concluded there was no alternative to using the bomb for military purposes during the war. The four scientists discussed the possibility of a technical demonstration, but determined that it would be unlikely to persuade the Japanese to surrender. They were also concerned that if Japan knew of a forthcoming attack, it might relocate its prisoners of war to the site of the demonstration explosion. The interim committee therefore recommended that the weapon be used against Japan as soon as possible, that it be used on a war plant surrounded by workers' homes, and that it be used without prior warning.[3] It is not known whether a different recommendation would have had an impact on the U.S. military command and on President Truman. According to Groves, Roosevelt would have even considered dropping the bomb on Germany as early as December 1944 if it had been ready at that time.[4] The decision to drop the bomb on a Japanese military target had been made at the top level in May. And so, on August 6 and August 9, 1945, the United States dropped two atomic bombs, one on Hiroshima and the other on Nagasaki.

Even after all these years, the jury is still out on whether it was the right decision, and what would have happened if the Americans had simply conducted a demonstration test. The atomic bombs dropped on Japanese civilian populations likely helped bring the war to a faster end, thereby sparing the lives of countless American and Japanese soldiers and civilians. But does that justify the horror of Hiroshima and Nagasaki? It is also true that some of the air raids during the war were even more deadly than the atomic bombs. The series of aerial bombings of Dresden between February 13 and 15, 1945, practically wiped out the city, the incendiary devices killing up to 300,000 people, according to some estimates. While Canada took part in the Dresden attack, did it also play a role in the bombing of Hiroshima and Nagasaki?

In the previous chapters, we have seen how Canada was involved in the development of the nuclear weapon project undertaken by the British (although it wasn't completed before the end of the war) and, indirectly, in the American project as well. But was Canada a direct participant, knowingly or otherwise, in the development and building of the atomic bombs that were dropped on Japan? And did Canada take part in the political decisions authorizing the bombings? The country could have played a material role in the Hiroshima and Nagasaki atomic bombs by providing uranium, plutonium, heavy water (used in the nuclear reactors that generated the plutonium), or polonium (used as a detonator in the plutonium bombs). Canada could also have played an intellectual role by assisting in the conception and research that led to the building of the bombs. And lastly, it could have played a political role by agreeing to the Hiroshima and Nagasaki bombings. Let's examine these points more closely.

The uranium in the Hiroshima bomb

The bomb that was dropped on Hiroshima contained uranium-235 that came from the Oak Ridge enrichment plant in Tennessee. But where did the natural uranium used at Oak Ridge come from? It could have been from Canada, because the mine in Port Radium, Northwest Territories, that belonged to Eldorado Gold Mines, was reopened in 1942 at the request of the Americans for their Manhattan Project. Unbeknownst to the Canadian and British authorities, Eldorado had signed a supply contract with the Americans. Once it was extracted, the uranium ore was refined at the Port Hope plant on the shores of Lake Ontario, then exported to the United States, where it was used by the Manhattan Project. But the Americans needed a huge amount of uranium, and the Canadian production wouldn't have been enough.

After a bit of digging, we discovered that the uranium used at Oak Ridge came from two different sources, neither of them in Canada. Most was from a stockpile of uranium that had actually been sitting in New York since 1940. It belonged to Edgar Sengier, a Belgian businessman and president of Union minière du Haut-Katanga which, in 1939, owned the biggest supply of uranium in the world. In December 1940, Sengier had secretly ordered over 1,000 metric tonnes of uranium ore to be shipped to a warehouse on Staten Island, for fear the precious resource would fall into the hands of the Germans. The uranium was from the Shinkolobwe Mine in the Belgian Congo, in what is now the Democratic Republic of Congo. When the United States launched the Manhattan Project, the American army acquired all of Sengier's ore.[5] Some of this uranium was used in the Oak Ridge enrichment plant. But before being shipped to Oak Ridge, the uranium had to be processed into uranium dioxide (UO_2). There were very few plants during the war that were producing UO_2 from uranium ore, and the Eldorado refining plant in Port Hope was one of them. It appears that part of the ore from the Belgian Congo was shipped to Port Hope before making its way to Oak Ridge.

The other source of the uranium used at Oak Ridge was the uranium recovered by the Alsos mission in Europe. Alsos was the codename of a joint operation by American intelligence agencies to uncover the military and scientific advances the Germans and Japanese had made during the war. Its main focus was German atomic research. When Germany was invaded by the Allied troops in the early months of 1945, a mission led by Colonel Boris Pash, a former Manhattan Project security agent, located the Germans' uranium stockpile (nearly 1,000 tonnes) in Stassfurt, in the Soviet-occupied zone. When the war broke out, the Congolese uranium had previously been stored in Belgium. Pash removed it right under the noses of the Soviets and had it shipped to England, then on to the United States. So, while the Canadian uranium was used by the Manhattan Project in various nuclear reactors, it seems it did not go into the construction of the Hiroshima bomb. Canada's participation appears to have been limited to the preparation of uranium dioxide from Belgian Congo that was destined for the Oak Ridge enrichment plant.

The heavy water and plutonium in the Nagasaki bomb

Heavy water was of paramount importance during the war because it could be used as a moderator in a nuclear reactor where, after absorbing a neutron, uranium-238 would transform into plutonium-239, which is fissile and can be used in a bomb. As we saw earlier, the Americans helped build a heavy water plant in Trail, British Columbia, and bought up its entire production. The heavy water produced in Trail was shipped by rail to Chicago, where it was used in experimental reactors,[6] not to produce plutonium.

The plutonium used in the bomb that was dropped on Nagasaki came from a reactor in Hanford, in the state of Washington, which was not a heavy-water reactor; rather, it was a graphite-moderated reactor similar to the first pile Fermi had built beneath the football stadium stands in Chicago. The information provided in the links on the B Reactor National Historic Landmark site (created by the Obama government in 2014) proudly claims that the plutonium used to make the bomb that was dropped on Nagasaki was produced at the Hanford site.[7] It therefore appears that the heavy water produced in Trail was not used in the reactors in which the Nagasaki bomb plutonium was made.

The polonium in the Nagasaki bomb

As we saw earlier, in November 1943, Bertrand Goldschmidt brought back to Montreal a considerable amount of polonium that he had prepared in New York at the request of the Americans. Once they collected it from him, they shipped the precious polonium straight to Los Alamos, where it would be used in atomic bomb research to detonate plutonium bombs. The polonium was considered an official contribution of the Montreal Laboratory to the Manhattan Project. Since their polonium requirements largely exceeded the amount isolated by Goldschmidt, and since polonium has a short half-life (138 days), in late 1943, the Americans established a research centre in Dayton, Ohio, to extract polonium from the uranium stocks they had in their possession.[8] Starting in 1944, bismuth was irradiated in a nuclear reactor at Oak Ridge.[9] When it absorbs a neutron, bismuth-209 becomes radioactive bismuth-210, which transforms into polonium-210 through a process called "beta decay." It was this polonium that was used as an initiator for the Nagasaki bomb. The polonium that was officially given by the Tube Alloys project to the Manhattan Project was used for research at Los Alamos, not as an initiator in the bomb that was dropped on Nagasaki.

Bomb conception and manufacture

One of the main differences between the Manhattan and Tube Alloys projects was that, starting in the summer of 1942, the American project was under the direction of the military, whereas, despite the outbreak of war, the Anglo-Canadian project remained a civilian-led enterprise. The Manhattan Project was directed by General Groves, together with lab director and physicist Robert Oppenheimer. In Canada, it was two civilians—C. D. Howe, the Minister of Munitions and Supply; and Chalmers J. Mackenzie, director of the National Research Council of Canada—who oversaw the Montreal Laboratory. As a result, and owing also to the fact that the American project was much further advanced than its Canadian counterpart, there was never any testing or research done in Montreal that was directly tied to the conception and building of a bomb. However, some of the British scientists who were "loaned" to the Los Alamos lab, including Otto Frisch and Rudolf Peierls, were involved in this phase of the Manhattan Project. Montreal's role focused on producing the material (plutonium) required to eventually make a bomb, but not on the actual conception or building of the bomb itself. The Canadian scientists and engineers who worked at the Montreal Laboratory did not therefore participate in the conception of the Hiroshima and Nagasaki bombs.

Political decisions

Prime Minister Mackenzie King took part in the Quebec Conference where an agreement on atomic research was reached between the Americans and the British. While Canada's head of government was the official host of the gathering, the Americans shut him out of the discussions on atomic weapons. And it is true that Canada committed funding to the project through the Montreal Laboratory, but its leaders had no say in the dropping of the bomb. The Quebec Agreement, which stipulated that the United States and the United Kingdom would not use the bomb "against third parties without each other's consent,"[10] made no mention of Canada. Mackenzie King refers, in passing, to "the use that will be made of the atomic bomb"[11] in his diary on July 26, 1945, only eleven days prior to the bombing of Hiroshima. At no time did the U.S. president seek out

Mackenzie King's opinion. Canada clearly played no direct role in the political decision to drop the bombs on the two Japanese cities. That said, Canada praised the bombings after the fact, as did Churchill's government in Britain.

This brief analysis leads us to conclude that Canada was not directly involved in the bombing of Hiroshima and Nagasaki, either through its natural, intellectual, or political resources. In other words, even if the Montreal Laboratory had not existed, the United States would still have dropped the bombs on Hiroshima and Nagasaki. However, Canada did play an indirect role by way of its uranium dioxide processing plant in Port Hope, its scientific know-how in preparing polonium, as well as its political support.

Proliferation

After the war, the Canadian government decided not to move forward with the development of a nuclear bomb. At that time, Canada was, after the United States and Britain, the third-most advanced country in terms of nuclear bomb development and, with the injection of massive financial resources, could have succeeded in building a plutonium bomb. Instead, the federal government opted to develop nuclear energy for the purpose of generating electricity and providing radioisotopes for use in cancer treatments.

But its noble intentions didn't mean Canada was as pure as driven snow when it came to nuclear bombs other than the two dropped on Hiroshima and Nagasaki. In the post-war period, Canada and some of the engineers and scientists who had worked at the Montreal Laboratory participated, even if only indirectly, in the military nuclear programs of the United States, England, France, India, and Israel. What's more, once the NRX reactor went critical, a plutonium production plant was commissioned in Chalk River and, between 1959 and 1964, part of the plutonium it produced (approximately 252 kg) was sold to the United States, where it was incorporated into that country's plutonium stocks destined for nuclear weapons.[12]

In 1954, Canada provided India with a copy of the NRX reactor known as CIRUS, for which the heavy water was supplied by the United States. The agreement with India stipulated that the reactor must only be used for peaceful purposes. But India nevertheless went ahead and used the spent fuel from the CIRUS reactor to develop its own plutonium reprocessing plant, employing that plutonium to conduct its first-ever nuclear test, codenamed "Smiling Buddha," in 1974.[13] Contrary to popular belief, the plutonium used in India's military program was not produced by its CANDU reactors (also sold to India by Canada), but rather by a replica of the NRX reactor.

Some of the scientists who had worked at the Montreal Laboratory also went on to play crucial roles in the development of Britain's and France's atomic bombs. In 1958, Bertrand Goldschmidt, for one, was actively involved in setting up, in the utmost secrecy, a plutonium plant in Marcoule, some thirty kilometres from Avignon, in the south of France. The plutonium produced there was used in "Gerboise bleue," the first nuclear bomb test conducted by the French in the Sahara Desert, on February 13, 1960, during the Algerian War. Similarly, several chemists and physicists who worked in Montreal helped build the Windscale plutonium plant, which provided the material used to detonate Britain's first nuclear bomb, on October 3, 1952, in a lagoon in the Monte Bello Islands off northwestern Australia.

Legacy of the Montreal Laboratory

Now that we have examined the military aspect of the Montreal Laboratory's legacy, let's take a look at its impact on the civilian world.

Academics

Before World War II, the physics departments at Canada's universities had only a handful of internationally recognized scientists. Subsequently, dozens of physicists and chemists from the Montreal Laboratory helped boost the ranks of academic faculties in Canada and around the world. As we saw earlier, Pierre Demers joined the department of physics at Université de Montréal, where he taught until the 1980s. George Volkoff, a physicist who worked in the theoretical physics division under Placzek in Montreal, returned to his teaching position at the University of British Columbia (UBC), where he helped build one of the country's leading physics departments, in part thanks to its particle accelerators.[14]

Several of the scientists at the Montreal Laboratory had been professors before the outbreak of war, but the hands-on experience they acquired at the Lab, and the interaction they had with other scientists there helped them make a name for themselves in scientific research circles worldwide. Without the Montreal Laboratory, the physics and chemistry faculties at Canada's universities would never have succeeded in attracting such large numbers of highly qualified professors in the post-war period.

Several American and French universities also benefited from the experience and expertise of the Montreal

hemist Fritz Paneth upon his departure from Montreal to go back to his teaching position at urham University, August 1945. On his right is Alma Chackett, and on his left, his son Heinz aneth. (Kenneth Chackett, personal archives of Alma Chackett.)

Laboratory researchers. Fritz Paneth, the director of the chemistry division, returned to Durham University in the U.K., where he headed up the radiochemistry department.

JEANNE LECAINE-AGNEW

Jeanne LeCaine-Agnew, a mathematician at the Montreal Laboratory who we discussed earlier in the book, went on to have a distinguished academic career in the United States. After the war, her husband, Theodore Agnew, obtained his PhD and was offered a teaching position at Oklahoma State University in 1948. Jeanne couldn't be hired at the same university because of the anti-nepotism laws prohibiting spouses from working at the same establishment. But in 1953, two circumstances worked in her favour: one was the significant increase in enrollment at the university (thanks to the scholarships offered by the U.S. government after the Korean War), and the other was the scarcity of professors. The dean of the mathematics department protested that it was unthinkable that he not be allowed to hire as a professor someone with a PhD in math from Harvard who lived right in Oklahoma City! Jeanne LeCaine-Agnew was eventually offered the position, and she went on to teach undergraduates and supervise master's and doctoral students, primarily in the area of number theory, until her retirement in 1984.

Owing to her work at the Montreal Laboratory, Jeanne LeCaine had more industrial experience than most of the mathematics professors in the United States at that time. So, she decided to launch a project connecting the space, nuclear, mining, and other industries to solve mathematical problems through research and application. Together with Robert Knapp, she co-authored the manual *Linear Algebra with Applications*, one of the first of its kind in this field to require a computer.[15] Professor LeCaine received numerous accolades, including the Outstanding Teacher Award in 1964 and 1978. On her retirement at the age of 67, she was named professor emeritus. Jeanne LeCaine-Agnew passed away on May 8, 2000, at the age of 83.

In 1946, the scientific committee of France's Commissariat à l'énergie atomique (CÉA) included among its members three former scientists from the Montreal Laboratory: Pierre Auger, Bertrand Goldschmidt, and Lew Kowarski. (Commissariat à l'énergie atomique.)

Research centres

Most of the scientists from the Montreal Laboratory continued their careers in nuclear research, either at Chalk River, Harwell in the United Kingdom, or France's Atomic Energy Commission (CÉA). After his time in Montreal and Chalk River, Lew Kowarski, for one, saw his career take off. He returned to France after the war, where he worked for CÉA under high commissioner Frédéric Joliot. In 1948, Kowarski headed up the group that designed and commissioned France's first nuclear reactor, Zoé, at Fort de Châtillon, south of Paris. The project was very similar to the ZEEP reactor. This earned Kowarski the singular distinction of having overseen the commissioning of the first nuclear reactor in two different countries! He was assisted in this endeavour by a number of scientists, including Jules Guéron and Bertrand Goldschmidt, from the Montreal Laboratory, as well as by Irène Curie. In 1952, Kowarski left CÉA for the European Council for Nuclear Research (CERN), where he became director of the data handling division. His first task consisted of supervising construction of a series of laboratories at a site near Geneva straddling the border between France and Switzerland. After retiring from CERN in 1972, Kowarski taught at Boston University and served as advisor to the United Nations. Lew Kowarski died in Geneva in 1979.

Where did Canada's nuclear program go from there?

After NRX went critical in 1947, the researchers and engineers at Canada's Chalk River facility helped design, build, and commission NRU, an even bigger research reactor than NRX, in 1957. The new reactor was a phenomenon in and of itself, remaining in operation for over 60 years. From 1974 to 2016, it produced the lion's share of the world's molybdenum-99, the most widely used radioactive isotope for imaging procedures in nuclear medicine.

NRU contributed to a host of research applications, including the investigation into the *Challenger* space shuttle accident in 1986: several steel parts of the shuttle rocket boosters underwent neutron stress scanning in the NRU reactor, and the findings helped prove that the parts were not to blame for the disaster. This is just one example of the research conducted over the course of more than seventy years at Chalk River, Canada's national laboratory.

Thanks to the research that was carried out in Montreal, Canada's nuclear program grew substantially in the 1960s.

The first electricity-producing nuclear reactor—the Nuclear Power Demonstrator—located in Rolphton, near Chalk River, Ontario, was commissioned in 1962. It was the first CANDU reactor (which stands for CANada Deuterium Uranium). Deuterium is the heavy hydrogen atom that makes up heavy water (D_2O). The use of natural uranium and heavy water by Canada's nuclear energy sector is one of the legacies of the Montreal Laboratory, allowing the country to avoid the uranium enrichment step although forcing it to rely on heavy water plants. It is extremely rare to find in Canada a sector in which the entire production chain is located on home soil, as is the case for the nuclear industry. Canada owns and operates uranium mines, uranium processing plants, nuclear power plants, radioisotope-producing reactors for medical purposes, as well as a host of intermediary suppliers.

While, in the beginning, the Montreal Laboratory was created to serve a military program, in the long term it helped launch a civilian nuclear industry. I am well aware of the controversy surrounding the civilian nuclear industry worldwide, especially after the Chernobyl and Fukushima accidents. My goal in writing this book is neither to promote nor condemn Canada's nuclear industry; rather, it is simply to tell the fascinating and little-known story of the Montreal Laboratory, and I hope I have succeeded in achieving that goal.

ACKNOWLEDGEMENTS

A whole lot of people helped me to research and write this book, and I will do my best to name each and every one of them here.

First, I owe a huge debt of gratitude to Sylvain Lumbroso, associate editor at Septentrion, who provided inestimable guidance, revised the original French version of the book in its entirety, and contributed directly to several chapters. Without him, my book would likely never have made it to print.

Thank you to Les éditions du Septentrion for the French version and to Baraka Books for the English version. Both publishers did remarkable work. I was very privileged that the book was translated by Katherine Hasting. Thank you to Judy Gibbs for the index; what she did is the modern equivalent of the Middle Age copyist.

I would also like to extend a special thank-you to Alma Chackett who, at the age of 102, still has an astonishingly quick mind. Her contribution to this book has been invaluable, particularly the many photographs dating from the war years. I corresponded extensively by email with her daughter, Daphne MacDonagh, who encouraged me tremendously and even conducted some research herself, notably on the passengers who crossed the Atlantic by ship during the war. The other Chackett sister, Lesley Wareing, read and commented an early version of this book.

Thank you to Irène Kowarski, daughter of Lew Kowarski and, in all likelihood, the last living person who sailed on the *SS Broompark* in June 1940, and who generously commented a preliminary draft of the book and sent me a copy of the film *Operation Swallow: The Battle for Heavy Water*. She also agreed to be interviewed for this book.

Thank you to Philippe Halban, who provided a copy of his father's diary, as well as photos and the remarkable postcard from Irène Joliot-Curie. He also agreed to be interviewed.

Thank you to Christopher Cockcroft, son of John Cockcroft, who directed me to the photo collection "Deep River 1945-1995" and the book *Cockcroft and the Atom*, which were of great help in drafting several chapters of my book, as well as for the photos of the Montreal Laboratory that he has carefully preserved over the years.

A huge thank-you to Janet Morgan, daughter of Frank Morgan, whose vivid sense of humour often made me smile, and who encouraged me at a crucial period in my research. She also commented the part of the book that refers to her father.

Thank you also to history buff Emmanuel Cortadellas for his expertise.

The following people reread and commented the parts of the book dealing with their parents: Hugh LeCaine-Agnew (son of Jeanne LeCaine-Agnew), Christopher Cockcroft, Thierry Leroux-Demers (son of Pierre Demers), and Janet and Christopher Mitchell (children of Joseph Stanley Mitchell).

The following people helped identify the employees of the chemistry division who appear on the photo on page 104:

Alma Chackett, Daphne MacDonagh, Neila Shumaker (daughter of Albert English), Anne Hardy (daughter of Geoffrey Wilkinson), Louise Grummitt (daughter of Bill Grummitt), Toni Leicester (daughter of Gerda Madgwick-Leicester), Janet Morgan, Christopher Vroom (son of Alan Vroom), Patricia Joan Chesney and Andrew Cook (children of Leslie Cook), Peter Martin (son of Graham Martin), Mark Yaffe (son of Leo Yaffe), Marg Loghrin (daughter of Allan Lloyd Thompson), Celia Steljes (daughter of Ethel Kerr), Maurice Guéron (son of Jules Guéron), Julian Betts (son of Robert Betts), Victoria Lister Carley (daughter of Maurice Lister), Michael Ornstein (son of Ruth Golfman), and Pauline and Peter Booker (children of Margaret Kingdon).

Sara Courant, another former employee of the Montreal Laboratory, helped me identify Anne Barbara Underhill in the photo on page 69.

Thank you to Claire Cohalan for her valuable comments, especially about medical isotopes.

I would also like to thank my father, Georges Sabourin, a former physics professor at Cégep Édouard-Montpetit, who read an early draft of this book. He inspired in me a love of science from my early childhood. My mother, Liliane, and my sons, Félix and Sylvestre, also provided encouragement and valuable feedback.

Lastly, I want to thank my wife, Claude Lefrançois, who threw her support behind this project from the outset, and who patiently listened to me read aloud the many emails I exchanged with the people mentioned above. Without her, this book would never have seen the light of day.

If the book contains any erroneous or inaccurate information, the fault is mine, and mine alone.

BIBLIOGRAPHY

AINLEY, Marianne Gosztonyi and MILLAR, Catherine. "A Select Few: Women and the National Research Council of Canada, 1916-1991," *Scientia Canadensis: Canadian Journal of the History of Science, Technology and Medicine* Vol. 15, No. 2 (41), 1991, p. 105-116.

ARCHIVES OF THE CANADIAN ARCHITECTURE COLLECTION. "Thomas E. Hodgson House (1892-1904)," cac.mcgill.ca, accessed October 31, 2019.

BARRETTE, Roger. *De Gaulle, les 75 déclarations qui ont marqué le Québec*, Quebec: Septentrion, 2019.

BEEVOR, Anthony. *The Second World War*, London: Weidenfeld & Nicolson, 2012.

BERNSTEIN, Jeremy. "A memorandum that changed the world," *American Journal of Physics*, Vol. 79, No. 440, 2011.

BOGART, Michelle and MEAD, Carol. "Archives Spotlight: The Jeanne Agnew Papers," *MAA Focus*, Mathematical Association of America, Vol. 28, No. 6, August/September 2008, p. 22-24 (accessed November 12, 2015), [online] www.maa.org.

BOTHWELL, Robert. *Nucleus: The History of Atomic Energy of Canada Limited*, Toronto: University of Toronto Press, 1988.

BRETSCHER, Egon, FRENCH Anthony Philip and MARTIN Elsie Beatrice Mabel. "Determination of the U-235 and U-238 fission cross section," Cavendish Laboratory Report B. R. 385, Cambridge University Churchill Archives Centre, CSAC 115.6.86, 1944.

BRODA, Paul. *Scientist Spies: A Memoir of my Three Parents and the Atomic Bomb*, Leicester: Matador, 2011.

BROWN, Andrew. *The Neutron and the Bomb: A Biography of Sir James Chadwick*, Oxford: Oxford University Press, 1997.

BRUSKIEWICH, Patrick. "George Volkoff and reactor physics in Canada," *Canadian Undergraduate Physics Journal*, Vol. VI, No. 3, 2008, p. 18-22.

BULLETIN OF THE ATOMIC SCIENTISTS. "A Report to the Secretary of War, June 1945," Vol. 1, No. 10, May 1, 1946.

CANADIAN JEWISH HERITAGE NETWORK. "The Joseph and Wolff Family Collection" (accessed September 20, 2016) [online] www.cjhn.ca.

CANADIAN NUCLEAR SAFETY COMMISSION. "Canada's historical role in developing nuclear weapons," May 28, 2012, available at nuclearsafety.gc.ca, accessed June 18, 2020.

CANADIAN WAR MUSEUM. "Life on the Homefront: Montréal, Quebec, a City at War," Ottawa, www.warmuseum.ca, accessed November 4, 2019.

CHACKETT, Gladys Alma and CHACKETT, Kenneth Frederick. "Methods of Estimating Fission Iodine in Irradiated Uranium Metal without Using Carrier," Montreal Internal Chemistry Report, CI-124, UK National Archives, reference AB 2/81, 1946.

CHACKETT, Kenneth Frederick and CHACKETT, Gladys Alma. "Report on the Analysis of Commercial Electrolytic Oxygen for Traces of Nitrogen," Montreal Internal Chemistry Report, CI-122, UK National Archives, reference AB 2/79, 1946.

CHURCHILL, Winston and a team of assistants. *The Second World War*, 6 volumes, Boston: Houghton Mifflin Harcourt, 1948-1953.

CLARKE, W. Brian, CROCKETT, James H., GILLESPIE, Ronald James, KROUSE, H. Roy, SHAW D. M. and SCHWARCZ, Henry P. "Henry George Thode, M. B. E. 10 September 1910 – 22 March 1997," *Biographical Memoirs of Fellows of the Royal Society*, London, Vol. 46, 2000, p. 500-514.

CLOSE, Frank. *Half Life: The Divided Life of Bruno Pontecorvo, Physicist or Spy*, New York: Basic Books, 2015.

DAHL, Per F. *Heavy Water and the Wartime Race for Nuclear Energy*, Institute of Physics Publishing, 1999.

DE GAULLE, Charles. *Mémoires de guerre*, 3 volumes, Paris: Plon, 1954-1959.

DEPARTMENT OF NATIONAL DEFENCE. "Early Defence Atomic Research in Canada," CRAD Report-4/79, Ottawa, Government of Canada, 1979.

DEPARTMENT OF RECONSTRUCTION. "Canada's Role in Atomic Bomb Drama," press release, Government of Canada, August 13, 1945.

DRÉVILLE, Jean. *Operation Swallow: The Battle for Heavy Water*, Trident and Hero Film, France-Norway, 1948 [film], 96 minutes.

DREYFUS, Claudia. "Still Charting Memory Depths," New York Times, May 20, 2013.

DUCKWORTH, Henry E. *One Version of the Facts: My Life in the Ivory Tower*, Winnipeg: University of Manitoba Press, 2000.

DUNWORTH, John Vernon and MITCHELL, Joseph Stanley. "Application of nuclear physics to medicine and biology," Health Division Internal Reports, HI-13, UK National Archives, reference AB 2/88, 1945.

DURHAM UNIVERSITY. "Graham Martin's Research Group" (accessed September 20, 2016) [online], chemistry-alumni.dur.ac.uk.

DUROCHER, René, LINTEAU, Paul-André and ROBERT, Jean-Claude. *Histoire du Québec contemporain, volume 1; De la Confédération à la crise (1867-1929)*, Montréal: Boréal, 1989.

DUROCHER, René, RICARD, François, LINTEAU, Paul-André and ROBERT, Jean-Claude. *Histoire du Québec contemporain, volume 2; Le Québec depuis 1930*, Montréal: Boréal, 1989.

DYSON, Freeman. "Nicholas Kemmer 7 December 1911 – 21 October 1998," *Biographical Memoirs of Fellows of the Royal Society*, Vol. 57, 2011, p. 189-204.

EGGLESTON, Wilfrid. *Canada's Nuclear Story*, Toronto: Clarke, Irwin and Company, 1965.

EMELÉUS, Harry Julius. "Friedrich Adolf Paneth 1887 – 1958," *Biographical Memoirs of Fellows of the Royal Society*, Vol. 6, 1960, p. 226-246.

ENGLISH, Albert C., CRANSHAW, Thomas E., DEMERS, Pierre, HARVEY, John A., HINCKS, Edward P., JELLEY, John V. and NUNN MAY, Alan. "The (4n+1) Radioactive Series," *Physical Review*, Vol. 72, No. 253, 1947, p. 253-254.

FEDORUK, Sylvia. "The Growth of Nuclear Medicine," *Proceedings of the 29th Annual Conference of the Canadian Nuclear Association and the 10th Conference of the Canadian Nuclear Society*, 1989, p. 3.

FERMI, Laura. *Atoms in the Family. My Life with Enrico Fermi*, Chicago: University of Chicago Press, 1954.

FREEMAN, Kerin. *The Civilian Bomb Disposing Earl*, Pen and Sword Books, 2015.

GAGNON-GUIMOND, Renée. "Leurs majestés au Québec. La visite royale de 1939," in *Cap-aux-Diamants*, Vol. 5, No. 4, Winter 1990, p. 26.

GASS, Henry. "Montreal in the age of the atom," in *The McGill Daily*, November 11, 2010.

GOLDSCHMIDT, Bertrand. "How it All Began in Canada – The Role of the French Scientists," *Proceedings of the Special Symposium: 50 Years of Nuclear Fission in Review*, Canadian Nuclear Society, 1989.

GOLDSCHMIDT, Bertrand. *Atomic Rivals* (trans. TEMMER, Georges M.), New Brunswick and London: Rutgers University Press, 1990.

GOLDSTEIN, Max, WALES Muriel and LODGE, Arthur. "Fast fissions in tubes: A numerical supplement to MT-199," *Montreal Theory Report*, MT-242, UK National Archives, reference AB 2/559, 1946.

GOWING, Margaret. *Britain and Atomic Energy 1939-1945*, Toronto: The Macmillan Company of Canada Limited, 1964.

GREEN, Malcolm Leslie Hodder and GRIFFITH, William. "Sir Geoffrey Wilkinson 14 July 1921 – 26 September 1996," *Biographical Memoirs of Fellows of the Royal Society*, Vol. 46, 2000, p. 594-606.

HARRIS, Eiran. "Interviews with Annette Wolff." Alex Dworkin Canadian Jewish Archives, Montreal, 1999, [audiotapes: cassette SC 1631, recorded July 30, 1999, and cassette SC 1632, recorded August 6, 1999], 60 minutes each.

HARTCUP, Guy and ALLIBONE, Thomas Edward. *Cockcroft and the Atom*, Boca Raton: CRC Press, 1984.

HAUKELID, Knut. *Skis Against the Atom: The Exciting, First Hand Account of Heroism and Daring Sabotage During Nazi Occupation of Norway*, Minot: North American Heritage Press, 1989.

HAYTER, Charles. "Tarnished Adornment: The Troubled History of Québec's Institut du Radium," *Canadian Bulletin of Medical History*, Vol. 20, 2003, p. 343-365.

HOFFMAN, Darleane C., GHIORSO, Albert and SEABORG, Glenn T. *Transuranium People, The Inside Story*, London: Imperial College Press, 2000.

HOWES, Ruth H. and HERZENBERG, Caroline C. *Their Day in the Sun: Women of the Manhattan Project*, Philadelphia: Temple University Press, 2003.

HURST, Donald Geoffrey. *Canada Enters the Nuclear Age*, Montreal: McGill-Queen's University Press, 1997.

JELLEY, Nick, MCDONALD, Arthur B. and ROBERTSON, R. G. Hamish. "The Sudbury Neutrino Observatory," *Annual Review of Nuclear and Particle Science*, Vol. 59, 2009, p. 431-465.

JONES, V. C. *Manhattan, the Army and the Atomic Bomb*, Washington: US Army Center of Military History, 1985.

JUNGK, Robert. *Brighter than a Thousand Suns*, New York: Harcourt Brace, 1958, p. 108.

KALBFLEISCH, John. "Wartime blackout drill caught Montreal off guard," *Montreal Gazette*, October 13, 1943.

KENWORTHY, J. M., BURT, T. P. and COX, N. J. "Durham University Observatory and its meteorological record," *Weather*, Vol. 62, No. 10, Royal Meteorological Society, October 2007.

KNIGHT, Amy. *How the Cold War Began: The Gouzenko Affair and the Hunt for Soviet Spies*, Toronto: McClelland and Stewart, 2005.

KOWARSKI, Lew. "Reflections on the Meaning of a Canadian 'First'," *Physics in Canada*, Vol. 32, No. 1, 1976, p. 4-6.

LAURENCE, George Craig. "Early Years of Nuclear Energy Research in Canada," Ottawa, Atomic Energy of Canada Limited, 1980.

LAURENCE, George Craig. "The Montreal Laboratory," lecture given at the annual dinner of the Canadian Association of Physicists, Sherbrooke, June 10, 1966.

LAURENCE, George Craig. "ZEEP – Canada's First Nuclear Reactor," *Physics in Canada*, Vol. 32, No. 1, 1976, p. 6-7.

LE TOURNEUX, Jean. "Rasetti à Laval," *Physics in Canada*, March/April 2000, p. 103-107.

LECAINE, Jeanne. "A Table of Integrals Involving En(x)," *Montreal Theory Report*, MT-131, UK National Archives, reference AB 2/536, 1945.

LECAINE, Jeanne. "Critical Radius of a Strongly Multiplying Sphere Surrounded by a Non-Multiplying Infinite Medium," *Montreal Theory Report*, MT-29, UK National Archives, reference AB 2/508, 1944.

LECAINE, Jeanne. "Milne's problem with capture, II," *Montreal Theory Report*, MT-119, UK National Archives, reference AB 2/534, 1945.

LECAINE-AGNEW, Jeanne and KNAPP, Robert C. *Linear Algebra with Applications*, 3rd edition, Salt Lake City: Brooks/Cole Publishing Company, 1988.

LINTEAU, Paul-André. *Histoire de Montréal depuis la Confédération*, 2nd edition, Montreal: Boréal, 2000.

MANHATTAN PROJECT NATIONAL HISTORIC LANDMARK. "The B Reactor National Historic Landmark," Hanford, Washington, available at manhattanprojectbreactor.hanford.gov, accessed June 18, 2020.

MANN, Anthony. *The Heroes of Telemark*, Benton Film Productions, United Kingdom, 1965 [film], 131 minutes.

MARTIN, Roy. *The Suffolk Golding Mission*, Southampton: Roy Martin & Lyle Craigie-Halckett, 2014.

MEADOWCROFT, Pat. "AECL Staff Hotels – List of Former Residents," Revision 21, August 2008, (accessed May 11, 2017) [online] bright-ideas-software.com.

MELFI, Theodore. *Hidden Figures*, Fox 2000 Pictures, 2016 [film], 127 minutes.

MELVIN, Joan. *Deep River 1945-1995*, Deep River: Jomel Publications, 1995.

MILNER, Peter Marshall. "Peter M. Milner, Society for Neuroscience" (accessed September 20, 2016) [online] www.sfn.org.

MITCHELL, Joseph Stanley. "Applications of recent advances in nuclear physics to medicine," *Health Division Internal Reports*, HI-15, UK National Archives, reference AB 2/205, 1945.

MITCHELL, Joseph Stanley. "Memorandum on some aspects of the biological action of radiations, with especial reference to tolerance problems," *Health Division Internal Reports*, HI-17, UK National Archives, reference AB 2/207, 1945.

MITCHELL, Joseph Stanley. "Provisional calculation of the tolerance dose of thermal neutrons," *Health Division Internal Reports*, HI-14, UK National Archives, reference AB 2/204, 1945.

MONTRÉAL-MATIN. "6o savants étrangers viennent s'établir à l'Université de Montréal pour poursuivre des recherches extrêmement importantes," January 8, 1943.

NATIONAL RESEARCH COUNCIL OF CANADA. *War History of Division of Chemistry*, Ottawa, 1949.

NÉMIROVSKY, Irène. *Suite française*, Paris: Éditions Denoël, 2004.

NUNN MAY, Alan. "Proposed use of a polymer pile at very small powers for the investigation of critical dimensions," PD-97, UK National Archives, reference AB 2/653, 1944.

OLIPHANT, Marcus Laurence Elwin, KINSEY, Bernard Bruno and RUTHERFORD, Ernest. "The Transmutation of Lithium by Protons and by Ions of the Heavy Isotope of Hydrogen," *Proceedings of the Royal Society of London. Series A, Containing Papers of a Mathematical and Physical Character*, London, Vol. 141, No. 845, 1933, p. 722-733.

OUELLETTE, Danielle. *Franco Rasetti, physicien et naturaliste (il a dit non à la bombe)*, Montreal: Guérin, 2000.

PAIS, Abraham. *Niels Bohr's Times: In Physics, Philosophy and Polity*, Oxford: Clarendon Press, 1991.

PALAYRET, Jean-Marie. Interview with Pierre Auger (in French), European University Institute, Historical Archives of the European Union, October 12, 1992 (accessed May 25, 2017) [online] archives. eui.eu.

PARR, Joy (ed.). *Still Running: Personal stories by Queen's women celebrating the fiftieth anniversary of the Marty Scholarship*, Kingston: Queen's University Press, 1987.

PERKOVICH, G. *India's Nuclear Bomb: The Impact on Global Proliferation*, Berkeley and Los Angeles: University of California Press, 1999.

PINAULT, Michel. *Frédéric Joliot-Curie*, Odile Jacob, 2000, p. 164-166.

PLACZEK, Georg and BLANCH, Gertrude. "The functions of En(x)," *Montreal Theory Report*, MT-1, UK National Archives, reference AB 2/496A, 1943.

QUIST, A. S. *A History of Classified Activities at Oak Ridge National Laboratory*, Oak Ridge National Laboratory, US Department of Energy, 2000.

REMARQUE, Erich Maria (trans. A.W. Wheen) *All Quiet on the Western Front*, New York: Little, Brown and Company, 1929.

RHODES, Richard. *The Making of the Atomic Bomb*, Cambridge: Cambridge University Press, 1986.

RICARD-CHÂTELAIN, B. "Quand l'histoire s'est écrite à Québec," *Le Soleil*, August 17, 2013.

ROMAN, Nancy Grace. "Anne Barbara Underhill (1920-2003), American Astronomical Society," 2003 (accessed November 12, 2015) [online] aas.org.

SCHROCK, Virgil E., SOMERTON, Wilbur H., WIEGEL, Robert L., and LAIRD, Alan D. K. "Mechanical Engineering: Berkeley" (accessed November 12, 2015) [online] texts.cdlib.org.

SEABORG, Glenn T. *Journal, 1946-1958*, Berkeley: Lawrence Berkeley University Laboratory, University of California, 1990-1991.

SPIERS, Frederick William. "William Valentine Mayneord 14 February 1902 – 10 August 1988," *Biographical Memoirs of Fellows of the Royal Society*, London, Vol. 37, 1991, p. 342-364.

SPINKS, John William Tranter. "Contamination in the active laboratories, Montreal Internal Chemistry Report," CI-73, UK National Archives, reference AB 2/45; 1944.

SPINKS, John William Tranter. *Two Blades of Grass: An Autobiography*, Saskatoon: Western Producer Prairie Books, 1980.

SUCCESSION PIERRE DEMERS. "Pierre Demers (1914-2017)" (accessed April 15, 2017) [online] centenairepierredemers.com.

TASCHEREAU, Robert and KELLOCK, Robert L. "The report of the Royal Commission Appointed under Order in Council P.C. 411 of February 5, 1946 to Investigate the Facts Relating to and the Circumstances Surrounding the Communication, by Public Officials and Other Persons in Positions of Trust of Secret and Confidential Information to Agents of a Foreign Power," Government of Canada, 1946.

TAYLOR, Hugh P. and CLAYTON, Robert N. "Samuel Epstein 1919-2001," *Biographical Memoir*, National Academy of Sciences, Washington, 2008.

U.K. NATIONAL ARCHIVES. "Friedrich Adolphus Paneth, 1940-1954," Records of the Security Service, file KV 2/2423.

U.K. NATIONAL ARCHIVES. "Hans Heinrich HALBAN: French; 1944 Jan. 01 – 1955 Dec. 31," Records of the Security Service, file KV 2/2422.

U.K. NATIONAL ARCHIVES. "List of Staff; National Research Council, Montreal Laboratory; 1943-1944," records of the United Kingdom Atomic Energy Authority and its predecessors, file AB 1/126.

U.K. NATIONAL ARCHIVES. "Organization of Engineering Division Chalk River and Montreal Laboratories – N.R.C.," files of the United Kingdom Atomic Energy Authority and its predecessors, file AB 2/123, 1946.

U.K. NATIONAL ARCHIVES. "Staff General "A" – T.A. Team; 1945-1946," records of the United Kingdom Atomic Energy Authority and its predecessors, file AB 1/187.

UNDERHILL, Anne Barbara. *The Early Type Stars*, Dordrecht: D. Reidel Publishing Company, 1966.

UNITED STATES ARMY CORPS OF ENGINEERS. "The Manhattan District History: Book III, The P-9 Project" 1947, declassified in 2005, available at www.osti.gov, accessed June 18, 2020.

UNITED STATES ARMY CORPS OF ENGINEERS. "The Manhattan District History: Book VIII, Los Alamos Project (Y) – Volume 3 Auxiliary Activities, Chapter 4, Dayton Project," 1948, declassified in 2013, available at www.osti.gov, accessed June 18, 2020.

UNITED STATES DEPARTMENT OF ENERGY. Human Radiation Experiments: ACHRE Report. Chapter 5: "The Manhattan District Experiments; The First Injection, Superintendent of Documents," Washington: U.S. Government Printing Office, 1998.

UNIVERSITY OF BIRMINGHAM. "The Nuffield Cyclotron at Birmingham" (accessed September 20, 2016) [online] www.np.ph. bham.ac.uk.

UNIVERSITY OF BRITISH COLUMBIA ARCHIVES. "Governor-General's Medal Awarded to Penticton Girl," *The Province*, May 14, 1941 (accessed November 12, 2015) [online] www.library.ubc.ca.

VERZUH, Ron. "Blaylock's Bomb: How a Small BC City Helped Create the World's First Weapon of Mass Destruction," *BC Studies*, No. 186, Summer 2015, p. 95-124.

VETERANS AFFAIRS CANADA. *The Battle of the Gulf of St. Lawrence*, Ottawa: Government of Canada, 2005.

VILLE DE MONTRÉAL. "Edgard Gariépy" (accessed April 15, 2017) [online], www2.ville.montreal.qc.ca.

VOLKOFF, George and LECAINE, Jeanne "Application of 'Synthetic' Kernels to the Study of Critical Conditions in a Multiplying Sphere with an Infinite Reflector," *Montreal Theory Report*, MT-30, UK National Archives, reference AB 2/509, 1944.

VON HALBAN, Hans, JOLIOT, Frédéric and KOWARSKI, Lew. "Liberations of Neutrons in the Nuclear Explosion of Uranium," *Nature*, Vol. 143, 1939, p. 470-471.

WALESONLINE. "Kenneth Chackett Obituary" (accessed September 20, 2016) [online] www.family-announcements.co.uk.

WALLACE, Philip Russell and LECAINE, Jeanne. "Elementary Approximation in the Theory of Neutron Diffusion," *Montreal Theory Report*, MT-12, UK National Archives, reference AB 2/499, 1943.

WALLACE, Philip Russell. "Atomic Energy in Canada: Personal Recollections of the Wartime Years," *Physics in Canada*, March/April 2000, p. 123-131.

WALLERSTEIN, Alex. "Would the atomic bomb have been used against Germany?" blog.nuclearsecrecy.com, accessed June 18, 2020.

WATSON, Peter. *Fallout. Conspiracy, Cover-Up, and the Deceitful Case for the Atom Bomb*. Public Affairs, New York: Hachette, 2018.

WEART, Spencer. *Scientists in Power*, Cambridge: Harvard University Press, 1979.

WERNER, M. M., MYERS, David K. and MORRISON, D. P. "Follow-Up of CRNL Employees Involved in the NRX Reactor Clean-Up," *AECL Report-7760*, Atomic Energy of Canada, Ottawa, 1982.

WHITEHEAD, M. A. "A Brief Survey of Science and Scientists at McGill," *Fontanus Monograph Series*, No. IX, Montreal, McGill University, 1996, p. 105-113.

WIENER, Charles. "Interview with Lew Kowarski." Niels Bohr Library and Archives, American Institute of Physics (accessed May 15, 2017) [online] www.aip.org.

WILLIAMS, Michael Maurice Rudolph. "The Development of Nuclear Reactor Theory in the Montreal Laboratory of the National Research Council of Canada (Division of Atomic Energy) 1943-1946," *Progress in Nuclear Energy*, Vol. 36, No. 3, 2000, p. 239-322.

NOTES

Origins of the Laboratory

1. Bertrand Goldschmidt, *Atomic Rivals*, Rutgers University Press, 1990, p. 164.

2. Danielle Ouellette, *Franco Rasetti, physicien et naturaliste*, Guérin, 2000, p. 68.

3. Renée Gagnon-Guimond, "Leurs majestés au Québec. La visite royale de 1939," in *Cap-aux-Diamants*, Vol. 5, No. 4, Hiver 1990, p. 26.

4. Roger Barrette, *De Gaulle, les 75 déclarations qui ont marqué le Québec*, Septentrion, 2019, p. 81.

5. Henry Gass, "Montreal in the age of the atom," in *The McGill Daily*, November 11, 2010.

6. "Thomas E. Hodgson House (1892-1904)," archives of the Canadian Architecture Collection, McGill University, cac.mcgill.ca, accessed October 31, 2019.

7. The house was torn down in 1976, along with numerous other heritage buildings, to make way for an apartment building, in preparation for the Olympic Games.

8. Henry Gass, op. cit.

9. The municipality located on the Island of Montreal bears the same name as the French city that was the site of one of the most ferocious battles of World War I.

10. Canadian War Museum, "Life on the Homefront: Montréal, Quebec, a City at War," www.warmuseum.ca, accessed November 4, 2019.

11. John Kalbfleisch, "Wartime blackout drill caught Montreal off guard," *Montreal Gazette*, October 13, 1943.

12. Interview with Alma Chackett, Swansea, Wales; September 9, 2019.

13. Per F. Dahl, *Heavy Water and the Wartime Race for Nuclear Energy*, Institute of Physics Publishing, 1999, p. 104-110.

14. Kerin Freeman, *The Civilian Bomb Disposing Earl*, Pen and Sword Books, 2015, p. 82.

15. Robert Jungk, *Brighter than a Thousand Suns*, Harcourt Brace, 1958, p. 108.

16. Michel Pinault, *Frédéric Joliot-Curie*, Odile Jacob, 2000, p. 164-166.

17. Margaret Gowing, *Britain and Atomic Energy 1939-1945*, Macmillan & Co, 1964, p. 51.

18. Jeremy Bernstein, "A Memorandum that Changed the World," *American Journal of Physics*, Vol. 79, No. 440, 2011.

19. Even the greatest physicists like Fermi and Bohr were skeptical at the time.

20. Margaret Gowing, op. cit., p. 109.

21. Interview with Alma Chackett, Swansea, Wales, September 9, 2019.

22. This was notably the case for John Cockcroft, Bernard Kinsey, and Alan Nunn May, who all previously worked on radar research.

23. At the time of Bretscher and Feather's suggestion, the elements with the atomic numbers 93 and 94 had no name. It was another Cavendish Lab physicist, Nicholas Kemmer, who proposed the names neptunium and plutonium, after the planets Neptune and Pluto, which are next in line after Uranus, moving away from the Sun (uranium has atomic number 92).

24. A minister in Churchill's War Cabinet (John Anderson) reached out to Malcolm MacDonald, the British Ambassador in Ottawa, to ask him to petition the Government of Canada in August 1942.

25. Report by Placzek and Volkoff (a Canadian physicist recruited by Laurence).

Laboratory at War

1. Glenn T. Seaborg, *Journal, 1946-1958*, Berkeley, Lawrence Berkeley National Laboratory, University of California, 1990-1991.

2 G. Laurence, "The Montreal Laboratory," talk given at the annual dinner of the Canadian Association of Physicists, Sherbrooke, June 10, 1966, p. 4.

3. Interview with Joan Wilkie-Heal, September 2019.

4. Margaret Gowing, op. cit., p. 184.

5. Bertrand Goldschmidt, *Atomic Rivals*, op. cit., p. 180.

6. B. Ricard-Châtelain, "Quand l'histoire s'est écrite à Québec," *Le Soleil*, August 17, 2013.

7. "National Research Council – Correspondence (1942-1943)," Library and Archives Canada, Fonds George C. Laurence, R15952-89-7-E.

8. G. Laurence, "The Montreal Laboratory," talk given at the annual dinner of the Canadian Association of Physicists, Sherbrooke, June 10, 1966.

9. Travel reports: F. A. Paneth, "Report on the visits to Chicago, Toronto and Hamilton, 2022 Oct. 1943," CI-23 ; "Report on the visit to Chicago on 8 Jan. 1944," AB 2/638 UK National Archives; R. E. Newell, "Visit to Argonne Forest – 9 Jan. 1944," AB 2/135 UK National Archives; H. G. Thode, "Chicago trip, 25-27 Jan. 1944," CI-38 ; B. W. Sargent, "Report on visit to Chicago 29 Feb. – 10 Mar. 1944,"AB 2/639, UK National Archives; A. N. May, "Visit to Chicago 13-27 April 1944," AB 2/650, UK National Archives; A. G. Maddock, "Chicago discussions of 5-8 June, 1944," CI-54; W. J. Arrol, "Report on a visit to Chicago 7-14 June, 1944," CI-49.

10. B. Goldschmidt, *Atomic Rivals*, op. cit., p. 207.

11. *Ibid.*

12. M. Gosztonyi Ainley, C. Millar, "A Select Few: Women and the National Research Council of Canada, 1916-1991," *Scientia Canadensis: Canadian Journal of the History of Science, Technology and Medicine* Vol. 15, No. 2 (41), 1991, p. 105–116.

13. J. Parr [ed.], *Still Running: Personal stories by Queen's women celebrating the fiftieth anniversary of the Marty Scholarship*, Queen's University Press, 1987.

14. P. Wallace, J. LeCaine, "Elementary Approximation in the Theory of Neutron Diffusion," MT-12, UK National Archives AB 2/499, 1943.

15. Interview with J. Wilkie-Heal, Southampton, September 2019.

New Era for the Laboratory

1. A. Nunn May, "Proposed use of a polymer pile at very small powers for the investigation of critical dimensions," PD-97, UK National Archives, AB 2/653, 1944 [note: "polymer" was the codename used by the Montreal Laboratory to refer to heavy water.].

2. Interview with Irène Kowarski, February 2020.

3. B. Goldschmidt, *Atomic Rivals*, op. cit., p. 211.

4. Interview with Lew Kowarski by Charles Weiner, October 20, 1969, American Institute of Physics, transcript: www.aip.org.

5. Interview with Lew Kowarski by Charles Weiner, op. cit.

6. B. Goldschmidt, *Atomic Rivals*, op. cit., p. 200.

7. J. W. T. Spinks, *Two Blades of Grass: An Autobiography*, Western Producer Prairie Books, 1980, p. 64.

8. B. Goldschmidt, *Atomic Rivals*, op. cit, p. 216.

9. "Staff General A," UK National Archives, AB 1/187, 1945, p. 73, "Tube Alloys – Canadian Team."

10. When Ken and Alma left England, they brought with them only their clothes. They weren't allowed to withdraw more than £10 from their bank in Great Britain, but when they arrived in Montreal, they found themselves a room to rent in a rooming house, and their expenses were covered by the authorities. A little while later, they moved into a pleasant apartment on Avenue Decelles, in a building where a number of other chemists from the Laboratory also lived.

11. "De la campagne à la ville, le Québec de 1910 à 1950 vu par Edgar Gariépy." City of Montreal Archives, www2.ville.montreal.qc.ca, accessed June 11, 2020.

12. J. W. T. Spinks, *Two Blades of Grass, An Autobiography*, op. cit., p. 65.

13. J. W. T. Spinks, "Contamination in the active laboratories," CI-73, UK National Archives, AB2/45, 1944.

14. J. W. T. Spinks, *Two Blades of Grass, An Autobiography*, op. cit., p. 65-66.

15. H. E. Duckworth, *One Version of the Facts: My Life in the Ivory Tower*, University of Manitoba Press, 2000, p. 80.

16. Personal communication (email) from Janet Morgan, July 17, 2016.

17. J. S. Mitchell, "Applications of recent advances in nuclear physics to medicine," HI-15, UK National Archives, AB 2/205, 1945.

A Den of Espionage

1. Amy Knight, *How the Cold War Began: The Gouzenko Affair and the Hunt for Soviet Spies*, McLelland and Stewart, 2005, p. 26.

2 Paul Broda, *Scientist Spies, A Memoir of My Three Parents and the Atom Bomb*, Matador, 2011.

3 "Friedrich Adolphus Paneth, 1940-1954," UK National Archives, Security Service, File KV 2/2423.

4. Laura Fermi, *Atoms in the Family. My Life with Enrico Fermi*, University of Chicago Press, 1954.

5. F. Close, *Half-Life, the Divided Life of Bruno Pontecorvo, Physicist or Spy*, Basic Books, 2015.

6. Ibid., p. 312.

7. R. Taschereau and R. L. Kellock, "The report of the Royal Commission Appointed under Order in Council P.C. 411 of February 5, 1946 to Investigate the Facts Relating to and the Circumstances Surrounding the Communication, by Public Officials and Other Persons in Positions

of Trust of Secret and Confidential Information to Agents of a Foreign Power." Government of Canada, 1946.

Conclusion

1. The document was signed by James Franck, Donald J. Hughes, James J. Nickson, Eugene Rabinowitch, Glenn Seaborg, Joyce C. Stearns, and Leo Szilard.

2. "A Report to the Secretary of War, June 1945," *Bulletin of the Atomic Scientists*, Vol. 1, No. 10, May 1, 1946.

3. M. Gowing, op. cit., p. 374.

4. Leslie R. Groves interview with Fred Freed, 1963, National Archives and Records Administration, RG 200, Box 4, "Groves, Leslie" quoted in Alex Wallerstein "Would the atomic bomb have been used against Germany?" blog.nuclearsecrecy.com (October 2013), accessed on June 18, 2020.

5. V. C. Jones, *Manhattan: The Army and the Atomic Bomb*, US Army Center of Military History, 1985, p. 64-65.

6. "Manhattan District History Book III, The P-9 Project," US Army Corps of Engineers, 1947, p. 5.32, declassified in 2005, available at www.osti.gov, accessed June 18, 2020.

7. "The B Reactor National Historic Landmark," Manhattan Project National Historical Park, Hanford, Washington, available at manhattanprojectbreactor.hanford.gov, accessed June 18, 2020.

8. "Manhattan District History Book VIII, Los Alamos Project (Y) – Volume 3 Auxiliary Activities, Chapter 4, Dayton Project," op. cit.

9. A. S. Quist, *A History of Classified Activities at Oak Ridge National Laboratory*, Oak Ridge National Laboratory, US Department of Energy, 2000, p. 13.

10. "Agreement Governing Collaboration between the Authorities of the U.S.A and the U.K. in the matter of Tube Alloys," https://www.atomicarchive.com/resources/documents/manhattan-project/quebec-agreement.html.

11. "The Diaries of William Lyon Mackenzie King," Library and Archives Canada, Thursday, July 26, 1945, p. 3, available at www.bac-lac.gc.ca, accessed June 18, 2020.

12. "Canada's historical role in developing nuclear weapons," Canadian Nuclear Safety Commission, May 28, 2012, available at nuclearsafety.gc.ca, accessed on June 18, 2020.

13. G. Perkovich, *India's Nuclear Bomb: The Impact on Global Proliferation*, University of California Press, 1999.

14. Patrick Bruskiewich. "George Volkoff and reactor physics in Canada," *Canadian Undergraduate Physics Journal*, Vol. VI, No. 3, 2008, p. 18-22.

15. Jeanne LeCaine-Agnew, Robert Knapp, *Linear Algebra with Applications*, 3rd edition, Brooks/Cole Publishing Company, 1988.

INDEX

Pontecorvo, Marianne, 136, 138–139
Port Hope, uranium refinement,
28, 158–159, 162
Prévost, Adrien, 71
"Provisional calculation of the
tolerance flux of thermal
neutrons" (Mitchell), 116
pubs use in route finding, 25

Quebec Conference, 59–65, 161–162

radar research, 31, 73–74, 81, 124,
186n22
radiation doses, 115–116
radiation sickness, 109, 110
radioactivity
care in working with, 51
radioactive decay chains, 68
Rutherford's research on, 16–17
See also Joliot-Curie Laboratory
radioisotopes. See isotopes
radium, 94, 112–113
Rasetti, Franco, 38–41, 39f, 138
Reynaud, Paul, 23
Rioux, Fernande, 42, 71
Robertson, Norman, 120
Roosevelt, Franklin D.
agreement signed by, 61, 71
Churchill's communication
with, 33
Manhattan Project launched
by, 33
at Quebec Conference, 59–61,
60f
views on atomic bomb, 156
Rose, Fred, 150–151
Rosenberg, Ethel and Julius, 144
Roux, Monseigneur le Recteur, 55
Russia. See USSR
Rutherford, Ernest
as Nobel Prize winner, 16–17, 18f

protégés/collaborators of
Chadwick, 27
Cockcroft, 25, 27, 79
Laurence, 42, 64
Oliphant, 27

Saclay nuclear research laboratory,
Halban at, 94
Sadler, Sheila, 110
security
decryption program, 144
implementation by Groves, 34,
73, 95, 129
of metallurgical lab, 126
spies targeting military/
scientific research, 121
Segrè, Emilio, 144
Sengier, Edgar, 157–158
Sereni, Emilio, 144
shoes confiscated as radioactive
waste, 108
"Smiling Buddha" nuclear test, 163
Smith, Phillip Dunford, 151
smog, nuclear energy as resolution
for, 81
Smyth Report, 131, 152
Spinks, John, 106–109, 107f
Spooner, Reginald, 131
Steacie, Edgar, 108
Stewart, John, 90
St-Laurent, Louis, 120
Strelioff, Mary, 106–109

theoretical physics research, 38,
48, 90
Thode, Harry, 64
Thomson, George, 124–125
Thornton, Robert, 144
Tizard mission, 32
triglycol dichloride (trigly), in
plutonium extraction, 97

ALSO FROM BARAKA BOOKS

The Einstein File, The FBI's Secret War
on the World's Most Famous Scientist
Fred Jerome

Montreal City of Secrets, Confederate Operations in Montreal
during the American Civil War
Barry Sheehy

The History of Montréal, The Story of a Great North American City
Paul-André Linteau

Mussolini Also Did a Lot of Good, The Spread of Historical Amnesia
Francesco Filippi

Patriots, Traitors and Empires, The Story of Korea's Struggle for Freedom
Stephen Gowans

Bigotry on Broadway, An Anthology Edited by
Ishmael Reed and Carla Blank

A Distinct Alien Race, The Untold Story of Franco-Americans
David Vermette

The Question of Separatism, Quebec and the Struggle Over Sovereignty
Jane Jacobs

Stolen Motherhood, Surrogacy and Made-to-Order Children
Maria De Koninck

Still Crying for Help, The Failure of Our Mental Healthcare Services
Sadia Messaili

MIX
Paper from
responsible sources
FSC® C100212

Printed by Imprimerie Gauvin
Gatineau, Québec